How to Not go Broke Ranching

Things I Learned the Hard Way in Fifty Years of Ranching

Walt Davis

Copyright © 2011 Walt Davis
All rights reserved.

ISBN: 1463611889
ISBN-13: 9781463611880
Library of Congress Control Number: 2011910478
CreateSpace, North Charleston, South Carolina

Acknowledgments

I need to acknowledge several people who have contributed to any success that I might have had as a rancher. The first would be my Dad, Willis Davis, who more than once told me, "I may not leave you anything else, but you will know how to work." He made good on his promise (although I still don't know why we had to sit on our horses at the back side of the pasture for forty-five minutes waiting for it to get light enough to start the gather).

Another man who had a strong effect on my young life was Monroe Hollowell who worked for Dad and was my boss when I became old enough to actually work. Monk was the first person to teach me to really see cattle, and he taught me fence building, windmill science, and cattle handling the way it should be done.

My life would have been much easier and more productive if I'd had the benefit early on of the counsel of Alan Savory, Stan Parsons, Bud Williams, Dick Divens, Fred

Provenza, and Ian Mitchell—Innes. I cannot emphasize enough the importance of the contributions of these men to the art and science of ranching.

05-07-2011
Walt Davis

Contents

Preface: How to Go Broke Ranching vii
    Without Hardly Trying

Chapter 1: Ranching in America 1

Chapter 2: Ranching Then-Now-Tomorrow 7

Chapter 3: Relationships between Grasslands 19
    and Grazing Animals

Chapter 4: The Evolution of a Rancher 31

Factors That Contribute to the Profitability and Stability of Ranching

Chapter 5: Get in Sync with Reality 47

Chapter 6: Maximize Biological Diversity 69

Chapter 7: Planned Grazing 91

Chapter 8: Substitute Management for Money 203

Chapter 9: Work with Rather Than against Nature . . . 227

Chapter 10: Stockmanship 239

Chapter 11: Correct Stocking Rate 251

Chapter 12: Correct Stocking Mix 255

Chapter 13: Plan for Drought 261

Chapter 14: Adapted Animals 275

Chapter 15: Short, Properly Timed Breeding Season . . . 283

Chapter 16: Adding Sheep or Goats? 289

Conclusions 305

PREFACE

How to Go Broke Ranching Without Hardly Trying

There have been how-to books written on lots of different subjects lately, and I thought that it might be nice to help those people who get up every morning and wonder, "What can I do today to lose money in the ranching business?" There is bound to be a huge demand for this kind of information since regardless of weather or markets, losing money is the main topic of conversation in every coffee shop in the country. The following is a collection of thoughts that should be useful in preventing any accidental outbreak of profitability.

Set your breeding season so that calves, lambs, kids, and the like, are born well before the onset of new growth. This ensures that the young animals will be big enough to utilize the forage when it arrives and thus will wean off heavy. You will hear some naysayers blather about the expense of maintaining lactating females without green forage, trouble getting females rebred, disease and death loss in the young, and even predictions that losses to predators will be worse because there are no young rats or rabbits around at this time of year. Some of this may be true, but we have got to have heavy weaning weights, right?

Always stock your country with just a few more head than you think you can handle. If you get lucky and have a good year, you will have more to sell, and you can't carry that grass over to next year, can you? Besides, the Fat Cow Feed Company has a new feed guaranteed to make stock do good on broom weeds and tree bark. No need to worry about how much grass you have; just book your feed early, and read the trade magazines to keep up with the latest feed supplements and mineral programs. You may hear tales about people claiming to winter with no supplemental feed but you know darn well it can't be done *here*. When you feed hay or other supplements, be sure to always feed in the same place so that you don't stomp out all your pasture. By feeding in the same place, you can get rid of the wasted hay and built up manure in the spring by setting it on fire. You won't have to worry about the fire getting away; there will not be anything to burn anywhere close to the feed ground. Just uphill from your stock pond is a good place for the feed ground so that the bare ground will cause more of the rain to run off and keep the pond full. Sure a little manure, urine and dirt gets in the water but the stock will still drink it.

Make all of your breeding decisions based on "what the trade demands" rather than on silly criteria such as the type of cattle that perform well under your conditions or animals with traits that allow you to reduce inputs. After all, who knows more about what is good for the industry—you or ConAgra? Along the same line, be sure to buy great-big, good old bulls that throw really heavy weaning calves. You may have to pull some calves, and you may even lose some cows—and after a few years, your cows may get pretty big and expensive to winter but even if they made four times as much money do you really want to have to tell people that your calves weaned at only 450 pounds?

Don't get talked into this management-intensive grazing stuff. Some people claim to be able to run more stock with better performance at less cost while improving their country, but you don't have time to do all that; besides, it won't work *here*. Stay away from these schools and seminars that claim to be able to teach you how to reduce your costs, increase production, and improve both your country and your quality of life. Even if what they teach really works, you will have to spend so much time thinking that you won't have time to rope more than twice a week.

Don't get caught up in the low-stress livestock handling like Bud Williams teaches. So what if it will reduce sickness and death loss, improve animal performance, cut labor costs, and lower your vet bills; you have to maintain an image, don't you? Do you really want your neighbors to see you **walking** behind a bunch of cattle? How are you going to have good horses and good dogs if they don't get lots of work? How can you teach a young horse cow sense with a bunch of cattle that never break a walk and all stay together? Just why are you in the cattle business, anyway?

Be sure to have the latest and best in hay equipment so that you can get your hay put up in a hurry. What is a few extra thousand dollars a year in interest compared to the satisfaction you get from all that shiny new iron? Plow up those old native hay meadows and plant one of the new "improved" varieties of annual hay plants. Sure, this means that you will have to till and fertilize, and plant every year, so that costs go up, and your country will erode some—but you will get bigger yields. Above all, don't let one of these nuts talk you into making hay out of surplus pasture or even into getting out of the hay business altogether by rationing out the standing hay with pasture subdivisions (or, even worse, temporary electric fence). This would mean that you would have to be thinking and planning all during the growing season about what kind of forage you can expect and what your forage demand will be all throughout the year. If ranchers didn't make hay all summer and feed it all winter, where would they get their caps?

For sure don't listen to listen to the "low -cost production advocates". So what if the price of beef and other commodities does tend to settle at the break-even price of the average producer so that only the low-cost producer can be profitable every year? What's the point of being a rancher if you can't drive a brand new four-wheel drive, four-door, dually pickup, ride really high-powered horses, and be known for having the most expensive and fattest purebred cattle in the county? Get to know all of the fertilizer, feed, vet supply, livestock equipment, and supplement salesmen so that you can keep up with the latest technology. Steer clear of taking advice from the old timers in the business. Just because somebody has made a living ranching for forty years doesn't mean they know the business. Most of those people don't even use embryo transfers or GPS-precision fertilizer systems.

Get those calves to market before they get too big. A three-hundred-pound calf will bring a lot more per pound than a five-hundred pounder. If you want to be able to brag about "topping the market", you better ship them early. There is a lot of talk nowadays about retained ownership instead of just shipping calves at weaning. Some people claim to dry winter their calf crop with just enough supplement to keep them healthy and growing normally and then make big and cheap gains on the spring flush. Even if they make three times as much net profit per calf, it is a lot of trouble to wean those calves and worry with them all winter. Don't believe the stories about people weaning calves in the pasture with no stress, weight loss, or sickness. You know darn well that won't work. Have a real market plan. Bankers don't like to loan money to buy when the market is down, so pay attention to the market and be ready to buy when prices go up. Bankers will always loan more money when the market is high; after all, they are the financial experts.

Most importantly, know why you are in the business and what you want to accomplish. There are some soft-headed nuts out there talking about how ranchers are not in the livestock business but rather are in the business of harvesting free solar energy by converting it first to biological energy (green plants) and then into wealth in the form of meat, milk, wool, wildlife, or whatever. You know what is really important, and there will never be a (pick one—Angus, white face, goat, sheep, other) on this place while you are running things. Your granddaddy was a _____ breeder, your daddy was a _____ breeder, and you are a _____ breeder—or at least you were until the bank sold you out.

In 1999, I wrote the above as a spoof on the foibles of the ranching industry. I don't apologize for poking fun at my fellow ranchers; in fifty-something years of ranching, I have made every one of these mistakes plus many more. To get some good out of all of the mistakes and missteps it seemed reasonable to put them in book form to get some good out of it all and maybe save some people the trouble of making these same mistakes all by themselves.

Introduction

I started thinking about writing this book when I began to do management consulting in 1974; thank goodness, I did not attempt to write it at that time. I had started consulting at the request of a banker friend who wanted me to help some of his struggling loan customers. The cattle business was in bad shape; the cattle cycle was at a periodic low point, and people were having trouble repaying the loans they had taken out when cattle were high. I was an ardent student of agricultural technology as it existed at the time, and I knew how to increase production of both crops and livestock. I firmly believed that "the solution to the problems of technology is more technology"; thus, I began to work with clients to improve their production by applying technology. This was when the green revolution was in full swing and all of academia and most commercial producers were certain that agriculture would never have another poor day just as soon as we figured out the right combination of weed sprays, hybrid plants, nitrogen fertilizer, crossbreeding, and insecticides for each producer.

At the time, I was recommending to clients the same techniques and practices that I was using, this stage of my consulting career came to an abrupt halt when I realized that neither I nor my clients were progressing toward becoming profitable. I don't think I was directly responsible for any bankruptcies, but the correct course was generally the exact opposite of the recommendations I was making at the time. Every practice that I recommended increased production; however, it also increased expenses on operations that were already drowning in red ink. In retrospect, I can see that many of the practices intended to remedy a situation actually made things worse. I finally began to realize that we should be managing not for production but for profitability and stability. In many cases, we were using practices that yielded short term gains in production but that caused long term reduction in the health of our soil, our forage, and our animals. True long-term profitability is possible only if the entire soil-plant-animal complex is healthy.

This realization started a process of study and experimentation that continues to this day. I don't claim to have all of the answers, but I now know that it is possible to manage the environment and to conduct agriculture in such a way as to benefit both. People around the world are changing their operations from the industrial model, which has been common since the early 1960s, to a model based on biological principles. This is not yet a majority position, but it is gaining converts, because it works from all standpoints: it is financially sound, ecologically sound, and it is a good way to live and raise a family.

CHAPTER 1

Ranching in America

This is not intended as a history of ranching; however, some background on the early days of ranching would be useful in explaining the conditions and attitudes that shaped the ranching industry. It has become fashionable in some quarters to bash ranching as the cause of all sorts of ills, both real and imaginary. In truth, bad things did happen to the land due to ranching; areas were overstocked, grasslands were damaged, and fragile land was plowed. To my mind, very little, if any, of this damage was malicious (ignorant, yes, but not intentional, as some would have us believe). I believe that most of the damage can be attributed to two causes: (1) a lack of understanding of the relationships between ecology and economics and (2) political ignorance. Sadly these two factors are still very strong influences today. When Europeans first began to venture out on the grass plains of North America, they were struck by two facts: (1) an unbelievable number of buffalo, elk, and pronghorns and (2) grass in amazing abundance. As with anything that is in great supply, neither the wildlife nor the grass was highly valued. Killing off the game was condoned as a means of forcing the plains Indians onto reservations, where they could be controlled, which would allow the opening of the area to civilization. When the buffalo herds were exterminated and the government began to bring in settlers, one of the conditions of receiving free land was that the grass must be plowed up and the land made "productive". The early stockmen were some of the few people who valued the grass; to most, it was a hindrance to progress, a wilderness to be subjugated and converted to productive use. Much of this attitude persisted and became part of conventional wisdom for many years. The idea that grasslands were something valuable that should be protected and improved is a relatively new concept that remained free of government involvement to any real extent until the early 1940s.

Ranching in the United States had its beginnings with the Spanish livestock operations in what would later become Texas, New Mexico, Arizona, and California. Originally, these operations were designed mainly to feed the soldiers, priests, and colonists who went forth to carry out the edicts of the Spanish crown to Christianize the natives (and, by the way, keep an eye out for gold and silver). All of the early Spanish explorers herded livestock along with them on their journeys as walking commissaries. Escapees from the horse herds of these explorers and from early Spanish ranches found horse heaven on the plains of Texas, and they were able to multiply—thanks to the Spanish custom of using intact stallions as saddle horses. (The Spanish caballero considered the gelding suitable only for ladies and priests and would as soon ride a burro as a gelding.)

In New Mexico and California, the local Indians more or less accepted Spanish rule (after some bloody pacification) and added animal husbandry to their ways of life. The immense distances from the outlying settlements back to civilization largely prevented the early ranches from becoming great commercial successes. Some livestock and some products such as wool and dried meat found their way back to Mexico overland, and hides and tallow were shipped from California by sea; nonetheless, growth of the ranches was stymied by a lack of markets. In Texas, southern New Mexico, and Arizona, the local Indians were less amenable to life as serfs on Spanish ranches but quickly adopted the horse and prospered, as the new ability to rapidly travel long distances made hunting buffalo easy and created lots of free time to devote to raiding and stealing more horses. The Comanche, in particular, turned raiding into an industry; they stole horses, cattle, and captives from New Mexico and sold to them in Louisiana. From Louisiana they stole to sell to be to the British and American traders to the north and they stole horses, cattle, and captives from Mexico and Texas to sell in New Mexico. They were equal opportunity operators that sold to whoever would buy and stole from whoever had what they wanted. They effectively halted all ranching and other agriculture in northern Mexico, western Texas, and much of what would become western Oklahoma, Kansas, and eastern Colorado for over a hundred years. They also were responsible for the point of this history lesson, —the thousands of horses and cattle that either escaped or were abandoned during this time and took up life in the wild. The cattle fared best in the brush country of southern Texas, because the buffalo (and more importantly, the buffalo wolf) seldom came into this area. This wild population would be the source of the Texas cattle industry that developed after the Civil War.

When the war ended, Texas was in terrible shape; many of its young men were dead, there was little industry or money, and the carpet baggers, with Yankee army help, disarmed the citizens and stole everything that wasn't red hot or immovable. The only commodities Texas had in quantity were wild cattle and wild horses, but even these had no value in Texas. They had to be captured and driven to the rapidly expanding population centers to the north to have value. It was this era that gave birth to the ranching industry and to the legendary American cowboy. It also fostered characteristics and attitudes in these early cattlemen that I like to think have been passed down through the generations.

My grandfather, Thomas Trammell, was born on the family plantation near Van Buren, Arkansas, on June 22, 1848, and was brought to Navarro County, Texas, by his parents in 1852. Phillip, his father, was a stockman and drover who regularly trailed both cattle and horses north into Missouri and east to New Orleans. Prior to the war, he would trail cattle by way of the Trammell Trace to Jefferson, Texas, where the cattle would be loaded on barges and towed to New Orleans.

Phillip had been orphaned early in life, but he became a very substantial man through his livestock operations. Later, the family lost nearly everything as a result of the war of northern aggression. In 1865, Phillip died on the trail to New Orleans, leaving Tom (my grandfather) to finish the drive, bring his father back to Texas, and to assume the responsibility of the care of his mother and eight siblings.

Although Granddad was able to spend less than three months total in school, he was determined to educate himself. He got a schoolteacher to scratch the alphabet on the skirt of his saddle so that he could learn it while riding and he taught himself to read; he died a very well educated man. Things were tough in Texas after the War; there was little money in circulation, the Comanche and Kiowa had driven the frontier back east over a hundred miles; killing, raping, stealing and taking captives to abuse and to sell. The Yankee carpetbaggers had disbanded the Texas Rangers; disarmed the few able bodied men left and were stealing everything in reach while sanctimoniously vowing to uphold the law. Granddad began rebuilding the family fortunes by gathering unbranded cattle that had been running wild since their owners had gone to war or had gone east to escape the marauding Indians. There were thousands of these cattle running wild that were considered to be worthless since, no local market for them existed. At age seventeen, Granddad and a cousin about his age would take two horses apiece, a sack of coffee, and a sack of salt, and go south into the brush country to catch wild cattle. They would live on coffee and fresh beef until they put together 250–300 head of cattle, which they would then drive to Indianola on the Gulf Coast, where they would be butchered for their hides and tallow. Later Granddad was to drive cattle back to Navarro County to build his own herds and also on to markets in the north.

My Grandfather died long before I was born, and Mose Newman, a nephew of Granddad's first wife, told me this story. I had quit telling it since it was obviously impossible for two men with no pens and only two horses each to gather and drive out of the brush a herd of wild cattle of this size. I broke my rule to tell the story to Bud Williams, perhaps the best stockman alive today, and his reaction was "So?" Probably Bud could do it but only if he had his wife, Eunice, to help. For the rest of us it is a case of realizing; that *"In those days, there were giants."*

Charles Goodnight, an early day's Indian scout and cattleman, personified the best of the traits that typified so many of the early cattlemen: self-reliance, honesty, loyalty, and a never-say-die attitude in all that they did. His partner, Oliver Loving, was killed by the Comanche as he and Goodnight returned from driving the first herd north over the Goodnight-Loving Trail. For the rest of his long life, Goodnight gave half of everything he earned to his dead partner's widow, not because of any contract but because he felt it was the right thing to do. The work was hard, the dangers were great, and the risks were horrendous, but these men and boys persevered and built a ranching industry that stretched from central Mexico to central Canada. These were hard men in a hard land. Today they are often characterized as bigoted, greedy, insensitive despoilers of paradise; I would suggest, however, that we all live in the world as it is during our time and not as we wish that it had been or would become.

CHAPTER 2

Ranching Then-Now-Tomorrow

Ranching developed as a way to convert low-value forage, often growing in remote or inaccessible areas, into valuable commodities that could transport themselves to market. At this stage, it was about as simple a process as could be conceived; let the animals eat the grass, allow the well-adapted ones to reproduce, walk some of the fat ones to market, let the grass regrow, and repeat the process annually. We have taken this very simple biological model and turned it into a much more complicated model that becomes more industrialized with each passing year. As late as the 1940s, most U.S. ranches were very low tech; water availability was improved with windmills or dirt tanks, and perhaps some supplemental feed was supplied, but basically, the livestock was expected to survive and produce on the natural production of the land. A high degree of natural selection occurred in that those animals that could not tolerate this regime did not reproduce.

Today most ranches are managed with a mind-set that is much more industrial than biological; modern conventional management looks at a ranch or a farm as an area of land where capital and labor are used to apply technology for the purpose of producing commodities. The natural processes that determine the health of the soil-plant-animal complex (water cycle, nutrient cycle, energy flow, and biological succession) are assumed to be completely controllable by technology and the state of their health to be of little consequence. Nowhere is this seen more strongly than in the decision several years ago of some soil-testing labs to stop testing for soil organic matter content based on the theory that nothing could be done to increase it and that its beneficial effects could be duplicated with technology. The era of modern agriculture is fifty to sixty years old, and its record is not good. We produce a lot more product but at a very severe financial and ecological cost. In 1940, we used one Calorie of fossil fuel to produce 2.3 Calories of food energy. Today, we use 10 Calories of fossil fuel to produce 1 Calorie of food energy. Modern agriculture does not have a very good track record when it comes to designing an agricultural system that is productive, profitable, and sustainable; one out of three doesn't cut it. We have wandered down one dead end after another, from confinement feeding of livestock, to replacing forages for ruminants with grain, to breeding "what the trade demands" or cattle that "will fit the box" instead of using practices and breeding animals that will produce profit and be sustainable under our local conditions. I suggest that we would be much better off both ecologically and financially if we looked at ranching, and indeed all agriculture, as being primarily about converting free solar energy into biological energy and then into wealth through photosynthesis, green plants, and animals.

Agriculture should be the art and science of promoting life so that we can harvest some of the surplus for our own use.

In size, purpose, and intensity, a ranch can range from a few acres of lifestyle ranch, where grandpa or grandma runs a handful of stock and keeps horses for the grandkids, to large-scale commercial ranches that are expected to produce a profitable return on investment. Regardless of the size or purpose of a ranch, grazing management—good, bad, or indifferent—determines the future condition of the soil-plant-animal complex that is the basis of that ranch. Unfortunately, this fact, an extremely important fact, is not completely accepted by many ranch managers and even some range scientists. People agree that rangelands have degraded, but too few understand the causes that have brought about the damage. Instead of looking at our management practices and trying to understand their total and long -term, overall effects, too often we blame past overgrazing, drought, and "invasive species".

Ranches tend to be located on marginal land in areas with erratic weather patterns, and under these circumstances, year effect in the form of extreme conditions (wet, dry, hot, or cold) have very strong and lasting effects, and it is easy to feel that our management attempts are futile in the face of nature's power. The amount and type of vegetation that can grow in an area is of course dependent upon on the local climate; thirty inches of rain per year and a long growing season will grow more grass than will fifteen inches and a short growing season. Yet it is very common to find well-managed ranches in fifteen-inch moisture belts growing far more grass than poorly managed ranches that receive twice the rainfall.

Most of today's degraded grassland is the direct result of poor grazing management; the cure for degraded grassland exists, however, it is using good grazing management to modify the factors that determine the health of grasslands. The proof that grassland degradation is not an unavoidable phenomenon is seen in two examples, one natural and one man-made. All of nature's destructive forces (drought, flood, fire, insect pests, and overstocking) had come and gone regularly for eons before humans became involved, but under nature's management, the grassland came back strong after each tribulation passed. Certainly there are areas today that have not made the comeback, but in all instances of which I am aware, humans have been the factor that prevented nature's healing. Only after man became involved could animals be held on land that could no longer feed and water them. Before modern man, when the grass was gone, the grazing animals left or died, which meant that neither the vegetation nor the soil was subjected to continuing abuse; with man supplying water and some amount of feed, animals were left on the land until both the vegetation and the soil were severely damaged.

Regardless of the size or purpose of a ranch, grazing management—good, bad, or indifferent—determines the future condition of the soil-plant-animal complex that is the basis of that ranch.

More easily seen are the positive effects of good land management; the exceptional ranch that always weathers misfortune better than its neighbors can be found in all areas. These ranches are not uncommon, but the understanding of why the ranches are exceptional is very uncommon; the normal reaction of neighboring ranchers is "Old Joe is sure lucky to have such a good ranch." Luck has nothing to do with it; Joe's ranch is good because Joe, and probably his daddy and his granddaddy before him, did what was needed to build and maintain a good ranch. Thoughtful ranchers all over the world have proven the effectiveness of good grazing management in restoring grasslands; the required management is well understood, and it can be monitored and fine-tuned to be effective in any area; a main reason that this is not done more frequently is our own impatience. We are geared to attack problems: spray this, plant that, blow diesel smoke on it, do something! Quite frequently, the ranch that receives rave reviews in all of the trade journals for *outstanding weed control* and *innovative brush eradication* will not be passed down to the next generation. Natural grasslands are extremely resilient and slow to change either for better or worse. Because of such resistance to change, the long-term results of management practices are sometimes hard to ascertain in the short term; one purpose of this book is to provide some tools that can be used to more quickly evaluate their effects.

Agriculture, ranching included, is the manipulation of resources—plants, animals, soil, water, sunlight, capital, labor, and knowledge—to create valued products. The products can be the usual food and fiber but may also include such things as development of wildlife habitat, improvement of watersheds, and increased health of the resources used in production. The complexity of the various components and their total interconnectivity guarantees that quick and simple answers tailored to one aspect of the operation without consideration of the effects on the whole operation are doomed to failure. To be valid, a practice must be sound ecologically, financially, and sociologically and must promote the well-being of all parts of the operation in question. Anything that is detrimental to one part of a functioning interrelated system will be detrimental to the whole system in the long term. Allan Savory was correct when he pointed out that it is impossible to manage land without considering the people and animals that depend on the land for their livelihood. Truly successful management must nurture all parts of the operation: people, soil, vegetation, livestock, wildlife, and finances. It becomes much easier to devise effective management programs when the manager has a good understanding of how these factors relate to each other and to the whole of the operation. I strongly recommend that all resource managers read *Holistic Management* by Alan Savory and Jody Butterfield to better understand these relationships.

Modern agricultural management theory is based on the concept that the success of an operation depends on its adoption of the latest and best management practices for dealing with each facet of the operation: weed control, animal nutrition supplementation, soil fertility, and so on. The hypothesis is that if each component of the production system is fully developed, the whole operation will function at its best; the problem, however, is that this is simplistic, linear thinking applied to a complex situation. If you read the research reports and advertisements in the trade journals and add up the research - proven 5 percent increase in weight gain from using brand X fly tag, the 7.5 percent increase from feeding brand Y supplement, and the 15 percent increase from using bulls from breed Z to all of the other increases promised by various practices you would conclude that thousand-pound weaned calves should be common and very profitable. The fallacy of this line of reasoning is that it fails to address the complexity and interconnectedness of the biological, financial, and sociological aspects of ranching. Although an action may be directed at only one facet of the complex, it will always have effects on all parts of the complex. These side effects may or may not be obvious: selecting for heavy weaning weights increases the weight of individual calves at weaning but also results in bigger cows that give more milk and require more feed; using an effective worming agent kills worms but also reduces the selection pressure imposed by the parasites and thus over time, it reduces the genetic resistance to worms in the host population, resulting in cattle that are ultimately less suited to their environment. Livestock losses to predation are reduced when wolves, eagles, coyotes, and other predators are eliminated, but this benefit may be overshadowed by the loss of the forage consumed by the increased numbers of jackrabbits, rats, and prairie dogs. In the 1940s and 50s, parts of west Texas were stripped of forage by an explosion of jackrabbits and rats when virtually all predators were eliminated through a combination of government and rancher predator control programs.

To how many jackrabbits and packrats do you assure long life when you shoot a coyote?

It is often difficult to predict the long-term effects of an action; the most effective manager is the one who can best predict the total and long-term, effects of available management practices and then utilize only those practices that promote the overall well-being of the operation. Many years ago, I attended a field day where an older rancher made the following comment: "Be very cautious when using practices that kill; make certain that the results you get are the ones that you want." At the time, I thought this was the silliest thing I had ever heard; why would I not want to kill weeds, internal parasites, horn flies, or coyotes? I did not say anything at the time, but years later, after having learned the hard way what he was trying to explain, I went back and thanked the rancher for his advice.

"First do no harm"; Good advice
For physicians and
For ranch managers

A practice that yields spectacular immediate benefits can be detrimental in its total effects. The use of highly available nitrogen fertilizer on pasture is a good example of this phenomenon; it brings about rapid improvement in growth and causes the forage to produce more tonnage with higher protein content. It also burns organic matter out of the soil and kills soil life, however, which greatly reduces the productivity of the soil. The prolonged use of high rates of nitrogen fertilizer brings about deficiencies of other mineral nutrients in the soil and in the forage produced on these soils. These deficiencies are passed on to the grazing animals and, combined with the lack of nutritional balance found in forage that has had its growth forced with nitrogen, can cause a wide range of maladies in animals. Impaired immune systems, bloat, metabolic upsets, retained placentas, eye problems, reduced fertility, foot problems, slow gains, and young born weak have all been traced to the effects of excessive use of nitrogen fertilizer. Excessive nitrogen fertilizer can also cause nitrate poisoning, which can be either acute and deadly or chronic and merely debilitating. In addition, the use of rapid-acting nitrogen fertilizer distorts the growth curve of plants into a boom and bust pattern, making it harder to provide grazeable forage over time, and requires that much of the forage be machine harvested at a higher cost. The purpose of this tirade is not to throw rocks at the fertilizer salesmen but rather to illustrate the complexity that managers must deal with on a daily basis. The successful manager will be constantly monitoring the effects of his actions and changing his management when he discovers faults.

The best manager is the one who can most accurately predict the total and long-term effects of the management practices used.

Nowhere does the law of unintended consequences play a larger role than in agriculture. Insecticides used to kill pest insects also kill the beneficial insects that prey on and control the numbers of the pest insect population. (Prey species always reproduce faster than their predator species; otherwise, the predators would starve. Thus, any reduction in the predator numbers gives a distinct advantage to the prey species.) Killing pecan casebearers with insecticides will also kill green lacewings and ladybugs that prey on plant aphids and will create an explosion of aphids. Insecticides also kill other species, such as soil dwellers, that were not even considered when the decision to use a pesticide was made. Some of these species die by direct action and some by disruption of the web of relationships upon which all life depends.

Herbicides kill beneficial as well as pest organisms, just as insecticides do, and they often have unforeseen side effects. For years, we have been told that glyposate herbicide is quickly denatured and rendered harmless by soil organisms; it turns out that the reason this material cannot be found in the soil is that it binds tightly to minerals such as calcium, potassium, magnesium, and others, masking its identity while rendering these elements unavailable to plants. There are reports that areas where this product has been used repeatedly show a marked reduction in plant growth and increased plant disease; this is another case of the remedy being worse than the disease.

Many "weeds" are excellent-quality forbs offering higher levels of mineral nutrients than are found in grasses. Forbs also contain secondary plant compounds such as tannin that can reduce animal maladies such as internal parasite infection and bloat. True weeds—those plants that have low value to grazing animals and are able to flourish under harsh growing conditions—still have value in the overall scheme. They are nature's way of preventing the waste of sunlight, moisture, and mineral nutrients that occurs when the soil surface is bare, the chemistry or biology of the soil is degraded, or poor grazing management prevents forages from thriving. The weeds are filling an ecological niche that, for whatever reason, is unfilled. By filling these empty niches, weeds prevent at least some of the wastage that would occur on bare ground, and by their life and death, they begin the process of healing the degraded areas.

The answer is not how to kill the weeds but rather how to change conditions so that forages can compete effectively with the weeds.

Weeds, pest insects, and disease organisms become troublesome only when conditions in the local environment favor their reproduction and survival and allow their numbers to explode. The real problem is not the "invasive weed" or the "virulent disease" but rather the conditions that give these organisms their competitive advantage. Weed and disease outbreaks are usually the direct result of poor management. Low biological diversity and weak immune systems brought about by stress of some sort are the usual culprits. We have developed a mind-set in agriculture that we must be in a constant war to the death with weeds, insects, coyotes, disease, and all other pest organisms. In reality, however, we would be much more successful if we could learn to manage for what we want rather than against what we don't want. The way to produce healthy forage swards, whether native range or alfalfa, is to provide the conditions that favor forage plants rather than the conditions that favor weeds and brush. Most of the time, good grazing management will be the most effective way to bring about the desired conditions.

Success is more likely when we manage
For what we want rather than
Against what we don't want

In the following chapters, I will lay out a series of changes that I believe can help move the business of ranching toward sustainable profitability while improving the resources on which it depends. To start this process, I need to define some terms and lay out some of the basic relationships between grasslands and grazing animals.

CHAPTER 3

Relationships between Grasslands and Grazing Animals

One of the most valuable skills that a stockman can possess is an understanding of the relationships between forage and grazing animals and the ability to use these relationships in management.

Contrary to the belief of some, grazing is not a win-lose situation; rather, it is beneficial for both forage and animals when done correctly. All of the great natural grasslands of the world developed in the presence of large numbers of grazing animals, so it is obvious that grazing per se is not damaging to grasslands. Conversely, it is also obvious that grasslands all over the world have been damaged and sometimes destroyed by grazing animals; given additional information, however, there are logical explanations for this apparent paradox. Conventional wisdom holds that the way to reduce damage to the land is to reduce the number of animals present. Yet, while this will reduce the amount of forage consumed, it is of no use in preserving the health of grasslands except in the case of a severely overstocked area. On natural grasslands, underuse is at least as damaging as overuse. The flaws in the destocking theory become obvious once certain facts about how forage plants grow and how animals graze are understood.

PROPER GRAZING

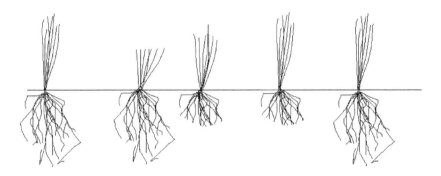

In the diagram above, the plant on the far left has a full complement of both leaves and roots, the second from the left has had leaves removed by grazing, the third plant has sacrificed roots to gain the energy needed to regrow the leaf area, the fourth figure has regrown leaves but not roots, and the last figure shows the plant fully recovered.

An important fact is that damage to plants by grazing occurs on a plant-by-plant basis. An area is not damaged by grazing; individual plants within an area are damaged. To illustrate this thought, consider what happens to a plant when it is grazed—first under good grazing management and then under successively poorer grazing management. In an ideal situation, a forage plant would grow a leaf, using energy stored either in the roots or in a seed, and this leaf would then convert solar energy to sugars and other starches through photosynthesis. These carbohydrates could then be used to fuel the growth of more plant material or stored for later use. If a cow comes along and bites the top out of the plant, no great harm is done; the plant simply uses some stored energy plus energy from the portion of the leaf that is left to grow more leaf area, and the cycle repeats itself. Because all animals, humans included, tend to eat the best first, the cow normally will move to another plant and not take a second bite farther down the same plant, as this part of the plant is tougher than and not as sweet as the top portion. If grass is scarce or the cow is particularly hungry, a second or even third bite may be taken from one plant. This is not ideal for either the cow or the forage plant, but it is not a disaster. The severely grazed plant will take longer to recover, because it has lost more leaves and must regrow them using mainly stored energy; if it is severely defoliated, it will sacrifice roots to gain energy, which will further slow its recovery. Thus, the growth rate of the severely grazed plant is slowed, but the plant has not been permanently damaged. The cow does not get as much nutrition per bite from the stems and older leaves as she would have gotten from younger material, but again, neither cow nor plant is severely or permanently damaged.

The serious damage to plants comes when plants are defoliated while they are growing using stored energy. If leaves are removed before they have had time to become functional and replace the energy required to grow them, the plant has no choice but to sacrifice more roots to get the energy to grow new leaves. This severely weakens the plant, and if repeated often enough, the plant will die.

Plants are overgrazed (severely damaged by grazing) only when they are grazed while growing on stored energy.

Overgrazing occurs in two ways: (1) if animals stay in one area long enough for previously grazed plants to regrow grazeable foliage or (2) if animals return to an area before the forage has had sufficient time to recover from earlier defoliation. The amount of time required for plants to recover is variable and depends on the growth rate of the plants.

The growth rate of plants depends on:
- the quality of growing conditions
- the amount of leaf area left after grazing
- the growth habits of the plant being grazed

Plants are overgrazed on a plant-by-plant basis. Overgrazing is a function of the frequency of defoliation.

OVERGRAZING

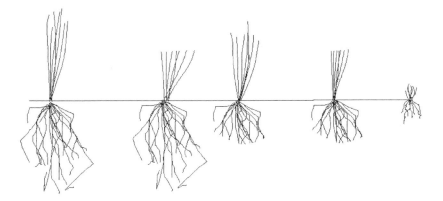

The diagram above illustrates what happens when a plant is grazed repeatedly without sufficient recovery time between defoliations.

As plants are damaged when they are grazed too often, a logical way to prevent damage should be to remove animals, leaving more plants per animal so that none is grazed too often. This would work except that to a grazing animal, forage quality is primarily a function of forage age. Immature broom sedge is better quality feed to a cow than overly mature alfalfa. As forage matures past a certain age, it becomes tougher, less digestible, and lower in protein and will be passed over by grazing animals in favor of younger forage. If grazing animals are continuously present, some plants will be grazed and regrazed as tender regrowth appears. Other plants of the same species will, for whatever reason, miss being grazed the first time and will become increasingly less valuable as forage and less efficient in converting solar energy to biological energy. In continuously grazed pastures, it is common to find plants dying from overgrazing and, a few feet away, plants of the same species suffering from underuse because old leaves are shading out the new growth. The high-quality native tall grasses are particularly subject to damage from underuse; this should not be surprising, as these plants evolved with grazing animals, and periodic defoliation was part of that association. Reducing stock numbers on a continuously grazed range does not stop overgrazing. Fewer plants will be overgrazed, but some plants will still be damaged by overgrazing while an increased number of plants will be damaged by underuse. Plants will not be killed or severely damaged in a single season of nonuse, but the growth rate of ungrazed plants will slow as the individual leaves age and lose effectiveness. If plants contain a large amount of dead stems and leaves over several seasons, they will begin to die back, starting in the most shaded area.

Forage plants require periodic defoliation to remain healthy.

Plants benefit from grazing by means other than defoliation; saliva from grazing animals supplies vitamins and enzymes that are useful to plants and the jerking, tearing motion of grazing triggers a beneficial effect in plant roots. When plant roots are disturbed through grazing, they respond by pumping sugars and some amino acids into the area of the soil that surrounds the hair roots of the plants. This seems like an illogical response; the plant has just lost part of its energy-producing structure, so why would it respond by sending energy into the soil? As nutrients are released into the immediate area of the plant hair roots, bacteria present in the soil use these nutrients to create a population explosion. Some bacteria can reproduce every twenty minutes when food is available, and a large number of bacteria are quickly produced in the area where sugars were released. Where large numbers of prey animals are present, predators will soon arrive. This is true of rabbits and coyotes, caribou and wolves, or bacteria and nematodes. As nematodes and other micropredators consume large numbers of bacteria, they take in more protein than they need and excrete the excess nitrogen as ammonia. This highly available form of nitrogen is released in the rhizosphere of the plant root hairs, where it can only be absorbed by the same plant that donated the sugars that started the process. The plant now has the nitrogen that it needs to regrow the leaves lost to grazing. Forage plants and grazing animals evolved together over eons of time and have developed a relationship that is beneficial to both.

It is a fact that when nature was running things, grasslands developed and thrived under heavy grazing pressure; it is also a fact that when man took over, many of the grasslands we managed began to deteriorate. What was different between the management schemes? Before man became involved with livestock, pack-hunting predators—man included—kept grazing animals grouped into compact herds. If an animal wandered too far away from the herd, it was eaten by lions, wolves, or some other predator. After a few generations of the dumb ones getting eaten, Cape buffalo and antelope in Africa, wild asses in Asia, and bison and elk in North America evolved into herding animals and lived in compact groups. Not shoulder to shoulder, except when they were actually fighting off attackers, but they grazed and lived in denser groups than cattle that live in a continuously grazed pasture do today. This bunching together of individuals determined how the animals grazed and moved. With many mouths to feed, the herd could not linger in one area very long and had to keep moving to fresh pasture. This same need for volume of forage also influenced where the herd grazed. Very young forage is tender and palatable but low in volume; old forage has more volume but is tough and slow to digest. Needing both volume and quality of forage, the herds tended to go to areas where the forage had recovered from the last defoliation and had good volume but had not yet become toughened by age. This is the stage of growth in which forage is most valuable to grazing animals, and it is also the stage of growth in which *growing* forage is least damaged by grazing.

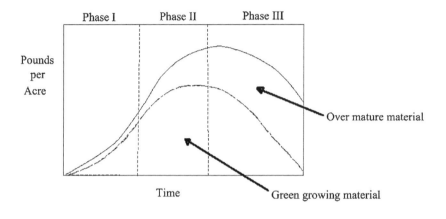

FORAGE GROWTH

Quality of forage to a grazing animal is primarily a function of the physiological age of forage.

Obviously, the animals could not always graze only this type of forage, but they sought out this stage of growth whenever they could and then moved on to fresh pasture. Given the choice, they would not come back to a grazed area until enough time had passed to dissipate the odor of their own manure and urine. This time lapse also allowed forage plants to recover from the effects of grazing and (just as important) time for soil life to break down and utilize the feast of nutrients provided in the tons per acre of body waste and trampled plants left in the wake of the herd. The trampling of an area grazed by a large, dense herd looks like the hind wheels of destruction, but in truth, high stock density is not damaging but beneficial to the plants and soil. Conversely, the effect of animals grazing continuously over the same area at low stock density brings a steady decline to both plants and soil.

High stock density grazing is the most powerful tool available to increase the health and stability of grassland.

The defining difference between high and low stock density is the recovery period; an area is hammered under high stock density, but then the animals move on, giving the plants time to recover, the soil time to rebound from trampling, and the soil life time to multiply and make use of the nutrients made available by the animals. When the density of grazing animals alternates between large numbers present and no animals present with adequate time between grazings, some very beneficial things happen at the soil level. With hoofed animals continuously present on an area, their repeated walking over the same areas creates a layer of compacted soil known as a hoof pan. This hoof pan slows the infiltration of both water and air into the soil and restricts the growth of plant roots. In areas that experience temperatures cold enough to freeze the soil a foot or so deep, this hoof pan forms to a much lesser degree, but in areas with milder winters, it can severely limit plant growth and soil health. The processes by which nature reverses soil compaction, other than freezing, are largely slow in action and easily reversed. The tunneling effect of the growth of plant roots, the swelling and contraction of soil particles as they take in and give up water, and the actions of soil life are the main processes that relieve soil compaction, and all are halted or reduced by the repeated compression of continuously present hoofed animals. If animals are concentrated and grazed over a number of areas, one area at a time, the healing processes have time to work, and soil compaction is greatly reduced. In this case, the action of a large number of hooves, or high stock density, acting on the soil for a short period of time is beneficial to the soil.

Continuous grazing is always damaging to grasslands to some extent.

The mass of organic matter from manure and trampled forage is incorporated into the soil by soil life, which greatly improves the water cycle and both the amount and rate of cycling of nutrients in the soil. This biological breakdown of organic matter releases carbon dioxide at the soil surface, and as the CO_2 is heavier than the surrounding air, the CO_2 concentration in the grass canopy increases. Plants take in the CO_2 they need for photosynthesis by opening stomata on the underside of their leaves, which also allows water vapor to escape; the longer the stomata are open, the more water is lost to transpiration. With the higher level of CO_2, the plants are able to take in the CO_2 they need in a shorter time, so water is conserved. Elevated CO_2 levels have little effect on climate and are a very good thing for plant growth; the real danger in CO_2 levels is that some brilliant idiot will come up with a way to drastically reduce the presence of CO_2 in the atmosphere and starve us to death by reducing plant growth.

Eons of time with millions of animals grazing at high stock density—a lot of animals in a small area for a short period of time followed by a recovery period—created the great natural grasslands. These grasslands had many different species of plants both grass and grass-like as well as forbs and other nongrass plants. They were home to many types of animals, from large to tiny, both above and below the soil surface. The areas where natural grasslands developed have erratic and extreme weather both within and between years; change (often radical) is the rule on these areas. Nature responds to these conditions by having different types of life available so that at least some things can flourish regardless of weather, wildfire, heavy grazing, or light grazing. The life and death of these multitudes of life forms, over time, created the deep rich soils that would later become the prime agricultural soils of the world. Other factors influenced the process, especially occasional fire and erratic weather, but grazing was a prime factor.

Proper grazing builds healthy grassland.
Improper grazing degrades grassland.

CHAPTER 4

The Evolution of a Rancher

In order to illustrate what I consider as flaws in the ways that ranches are and have been managed; I am going to use my personal experiences. Any resemblance to dumb things that you have done is purely coincidental.

In the words of the great philosopher, Pogo, "We have met the enemy, and he is us."

My people has been involved with livestock for many years, with one branch of the family turning out successful stockmen for at least five generations. There have been scurrilous rumors that some of my ancestors in Scotland and later in what would become Texas were not too concerned with the origin of the stock that they drove to market, but I feel certain that there were good explanations for any English or Mexican brands that were mixed in the bunch.

I grew up on a small ranch in west Texas that was run in much the same manner that my grandfather and great-grandfather had operated.

Wild Cows and Wild Cowboys

In the early 1940s my Dad, Willis Davis, and Walter Boothe were partners on the gyp lease and had it stocked with 200 Hereford cows. Those seven sections was as the saying goes "a dammed good place to lose a cow," the brush was 10-12 feet tall and so thick that anywhere you could see 100 feet was considered a clearing. It didn't take long for those old Hereford cows to realize that they could get through the brush a lot faster than a man could horse-back and that if they didn't want to go, he couldn't make them. Screwworms were still bad at this time and stock infected with this pest had to be treated or they would die. The constant chousing caused by the need to rope and doctor "wormies" didn't help make gentle cattle. One or two men could doctor wormies with enough riding, mainly because the horses could smell the worms and would go to an infected animal like a bird dog goes to quail, but it took a lot of help to pen the cattle. The world was at war and many of the countries' young men were in uniform. The draft board issued exemptions for men deemed essential to food production but even without this exemption cowboys would still have been in fair supply. There were lots of boys too young for the army but with 5-10 years' experience cowboying and quite a few of the older hands were too broke up to pass an army physical. This was a time and an area with lots of farms and small ranches; the people still lived on and made their living from the land. There were schools at places like Maryneal, Palava and Eskota and help was available to gather a pasture, work calves or to do whatever needed doing horseback. It was a little harder to find help to build fence or drench sheep. It has been a long time but the men that I remember being at one or more workings on the gyp lease were Dad, Mr. Walter Boothe, Ray Boothe, Spot Hale, Alvin Estes, Walter Estes, Monk Hollowell, Whop Smith, Charlie Smith and Blunt Carson. At five years of age my first "cowboying job" was also on the gyp lease at a calf working. I was in charge of dragging up wood and tending the branding fire. I wasn't allowed to ride my own horse during the gather, too easy to get lost when the run started. Besides even the steady old horses that I got to ride would go to the sound of a cow popping brush like kids to the bell on an ice cream wagon and charge any hole in the brush big enough for them to fit through, anybody on their backs had to look out for his own self. I remember the sensation of riding behind Dad on cow chases with a saddle string clenched in each hand and my face buried in his back. The sudden jerks, swerves and jumps would put a carnival ride to shame and it was even more gut wrenching because I couldn't see them coming. Dad's old high cantle saddle and bat wing chaps protected my legs and his body kept most of the limb whips off me even when he was stretched out flat on the horses' neck so we would fit through a tight hole. I learned quick to mimic every sway and bend that Dad made to keep him between the brush and me. The sounds and smells are still vivid; horses grunting, brush popping and men cussing mixed with the smell of horse sweat, man sweat, dust and Lucky Strike tobacco. This was my introduction to the world of men and to the fact that there were some things that we didn't do and some things that we didn't talk about around the womenfolk. I knew that we cut the testicles off of the bull calves so that they wouldn't breed the heifers long before I knew why they would want to breed the heifers. It was hard dangerous work and some of the men were a little rough around the edges but the coarse language and violent horseplay stopped the moment a woman came into view. It was a time when right and wrong were still absolutes and everyone was expected to, and most did, live accordingly. It was about this time too when Dad impressed on me that both his and my first duty as men was to take care of and protect our womenfolk in every way. The political correctness police would have a hemorrhage today but it was heady stuff for a kid trying to learn to be a man.

WWD *The Gyp Lease Tales and Other Lies*

Grazing was more or less continuous, with young being born mostly from mid to late spring, and the livestock was expected to obtain essentially all of their nutrition from the land as grazed forage. Salt and minerals were supplied, but no hay or other winter supplementation was provided. The exception to this was when cows were bred to calve in the winter to avoid screwworm infestations in the baby calves; these cows were given a little protein in the form of cottonseed cake. The heaviest workload was finding and doctoring screwworm cases. We spent many long days riding to look for and doctor screwworm cases to avoid losing animals; someone had to see every head of stock at least every three or four days during screwworm season, which started in April and lasted until we had a killing frost. About the only upside to this was that we got to be pretty good hands with a rope, and our horses got lots of work.

Screwworms

The screwworm was, and in parts of South America still is one of God's more unpleasant creatures. The adult female fly lays her eggs in open wounds on all kinds of animals where the eggs hatch and the resulting worms proceed to eat the host animal alive. Unlike other fly maggots, which only eat dead flesh and can actually be beneficial to the host animal, the screwworm eats only live flesh. Prior to screwworm eradication, wildlife experts say that half or more of each year's fawn crop in the south would be killed by screwworms. The adult flies lay eggs on the fresh navel wound and any calf born in the peak of screwworm season was almost certain to become infected. The mother cows, deer and other animals did their best to lick their babies clean but often they failed. Not only baby animals were at risk; an open wound of any kind, a thorn scratch or any break in the skin and all animals including humans could be infected. Ranchers spent a lot of time and money trying to protect their animals with only some very crude insecticides to work with. The momma cows looked upon the medicine we put on their calves as the problem and proceeded to lick it off as soon as possible. We tried all sort of recipes from coal tar to creosote to formaldehyde to try to keep the cows from licking the worm dope off the calves; the cows would grimace and slobber but they wouldn't quit licking until they got all that nasty stuff off their baby. I got an education on the degree of dedication this took when the big steer whose head wound I was doctoring while straddling the chute above him butted the bucket of worm dope into my crotch. I jumped down, shed my clothes in record time, jumped into the water trough and still wound up with a hell of a lot of blistered hide in some very uncomfortable places. Eradicating the screwworm in the U.S. and clear down to Panama was one of the few government programs ever that actually worked.

To this day if I get caught in a rain, my saddle smells like Smear 62 and EQ 335 and it has been close to 40 years since I used either kind of worm dope to doctor a case of screwworms. About that long ago, I was doctoring wormies in the corral on the Oklahoma ranch with Dad and Edger McCleary when Bill, our neighbor to the east, drove up and fell in to help. We had a squeeze chute and crowd pen so we could put each animal in the chute and treat its wound without having to wrestle with them. I had a little wooden paddle that I used to dip the worm dope up out of the jar and smear it into the wound but Bill didn't bother with that, he used his fingers.

Cow people don't tend to be too fastidious about cattle manure, blood and other bodily secretions. The old saw is you can't claim to be a cowboy till you've eaten five pounds of cow manure and drunk a quart of cow pee. So Bill using his bare fingers didn't turn any heads but several stomachs got a little queasy when Bill decided he needed a fresh chew of tobacco, slung off most of what was on his fingers and dug the old chew out of his jaw without even hunting a rag to wipe his hand.

Screwworms dictated everything we did when they were bad; we tried calving in the wintertime though this was hard on the cows and we worked calves on the hottest days of the year hoping the wounds would dry and form a crust before flies could lay eggs. Someone had to ride constantly from late April until a killing freeze doctoring wormies; a lot of the country was brush covered and infected animals tended to hide so we relied a lot on our sense of smell and the sense of smell of our horses to find them. I have ridden horses that could smell screwworms from a quarter mile away and go as straight to the infected animal as any good bird dog could go to quail. I doubt that anyone who ever smelled a case of screwworms ever forgot the stench. If we missed an animal with an infected wound, the animal died a horrible death.

Sometime in the early 1960s I went with Sam Lambert to look at a ranch he was thinking of buying in southwest Colorado. That was one of the roughest ranches I have ever been on; the house was built on the most level spot I saw in a long day of riding and when they leveled the floor, they had room to park their vehicles underneath on the downhill side. The man that owned the ranch was originally from Snyder Texas and when we asked how they came to be there, he said "Screwworms, I left Snyder with a bunch of screwworms in a bottle and when I got to where nobody knew what they were, I bought a ranch." Anybody who ranched in the South when screwworms were bad would understand.

WWD *The Gyp Lease Tales and Other Lies*

After the screwworm was eradicated in the 1960s, we stopped winter calving, which eliminated the need for winter supplementation. Dad usually had some stored feed of some sort available in case of ice storms, but these were rare, and the feed was seldom used except to feed horses or other stock left up in the corral. I remember a stack of bundle feed that remained in a fenced stack lot until it melted into the ground. As was common in the area at the time, Dad ran cattle, sheep, and goats all together and thus made use of the full range of forage available. On most commercial ranches, animals were selected based on how well they performed under the existing conditions and so tended to be easy fleshing, moderate in size, and well adapted to the style of management. Cattle were smaller than is common today and beef cows didn't give a lot of milk. The story is told about one west Texas rancher whose wife didn't have enough milk for their new baby; he had to rope and milk two or three range cows a day to keep the six pound baby full. For many years there was no feasible way to provide range animals with supplements beyond salt and minerals so animals that required more inputs were actively selected against. Production decisions were made based on the resources provided by the land. It was not the ranch's job to provide what the livestock needed; rather, it was the livestock's job to produce on what the ranch provided. Ranching in this manner was profitable and so long as stocking rates were kept reasonable, damage to the grasslands occurred very slowly. Continuous grazing always degrades grasslands, but had this management style been maintained with the addition of some form of time-controlled grazing, both profitability and sustainability could have followed.

It was not the ranch's job to provide what the livestock needed; rather, it was the livestock's job to produce on what the ranch provided

---

I went off to college in 1956 to study animal husbandry and soon learned how poorly my ancestors had been doing things for the last two hundred years or so. In December 1961, I left Texas A&M College after five and a half years of undergraduate and graduate work in animal husbandry and animal breeding and took over the management of the family ranch in southeast Oklahoma. At this point, fresh out of college, there was very little that I did not know about the ranching business. This golden period of enlightenment lasted maybe six months, and then my real education began.

I was armed with all of the latest technology and primed to take Davis Ranch into the glorious future. I kept up with faithfully and implemented all the new technology and instigated the best possible programs for soil fertility, weed control, animal nutrition supplementation, animal health, pecan pest control, and pasture improvement. I started an AI program and a performance-testing program to increase weaning weights and moved our calving date to January, as research proved that calves born in this month had the highest weaning weights. Results were immediate and spectacular, and production increased for both mother cows and stockers on both per-head and per-acre basis. Pecan production went up, as did both hay and grain crop yields, and we were taking in substantially more money in sales. Not everything was sweetness and light, however; costs kept going up, as it took more and more inputs to just maintain the status quo, and our operating loan note kept getting larger. It seemed that every time we sprayed to control one pecan pest, another would immediately take its place and demand control. Results that we had obtained in the past with fifty pounds of nitrogen fertilizer now required a hundred pounds plus some potassium. Weeds were a bigger problem than they had been when we used no weed spray and required more expensive materials for their control. As our cows became larger due to our selection for heavier weaning weights, we could run fewer of them, and they required much more supplementation. We also began to have more problems such as hard births, poor breed back, retained placentas, foot rot, and eye problems. I followed university recommendations on animal health; we wormed the cowherd twice a year, vaccinated for five or six different diseases, kept out a medicated creep feed for the calves, back poured all cattle in the fall with organic phosphate insecticide, fed mineral medicated with antibiotics and vitamins, and sprayed the cattle every 25–28 days from April to October to control horn flies. I carried a vial of epinephrine in my saddle pocket to treat the occasional animal that could not tolerate the high dosage of poison; there is no telling what I cost us in animal stress and poor production. Aside from the cost and the stress, my practices made it impossible to identify animals with genetic weaknesses and remove them from the herd.

In retrospect, it is easy to see what was happening, but we did not see it at the time. By destroying the organic matter content and life in our soil with acid salt fertilizers and tillage, we destroyed nature's ability to maintain the productive capacity of the soil, which led to reductions in both the quantity and quality of forage. The nitrogen fertilizer that we used so freely destroyed the methods nature uses to supply nitrogen to plants and left us ever more dependent on fertilizer. Dr. Bill Knight of Mississippi State pointed out this danger when he commented, "Sixteen pounds of actual nitrogen will kill every rhizobia bacterium on an acre of sandy soil." We destroyed the beneficial insects that preyed upon the pecan pests and animal parasites as we poisoned the pests; the situation we created was similar to what happened when we killed off all the coyotes, bobcats, and foxes and let the jackrabbit numbers explode. In our efforts to create "clean" monocultures of Bermuda, fescue, and other crops, we destroyed the biodiversity of our vegetation and created ecological "empty niches" that nature filled with a weed explosion. In using broadleaf poisons, we took out an entire class of plants, losing their abilities to concentrate certain minerals and their unique nutrient compounds. By selecting for larger cattle able to eat more corn in the feedlot and produce a carcass that "fits the box," we created cattle that were unable to produce as ruminants should on forage alone. We bred the ability to fatten easily out of our cattle to satisfy the packers' demand for less back fat and in so doing, we destroyed the cattle's ability to store nutrients as fat during good conditions and use this stored fat to survive the bad times. We had created "a pig in a cow suit" that required high levels of expensive inputs to survive and in doing so we destroyed our one economic advantage of cheap gains on grass. The result was a management nightmare and economic disaster that we were forced to recognize by the break of the cattle market in 1974; our choices were to reduce expenses or to go bankrupt.

When it became so painfully apparent that the cost of production had to come down, the obvious place to start was where expenses were highest. The farming operation, intended to provide year-round quality grazing for the cattle, was phased out, greatly reducing machinery, fuel, and labor costs. Nitrogen fertilizer was replaced with forage legumes, which also reduced weed pressure. Adjusting calving season into the spring and thus more closely matching nutritional demand to forage production reduced supplemental feed costs, reduced calf losses, and improved fertility. As the late-born calves were lighter, we began to carry them through the winter and sell them as heavy feeders the next summer after taking fast and cheap gains during the spring flush. As an aside, although it was not a conscious decision, our weaning weights declined over one hundred pounds from their high point when we were losing money to what they were when we became consistently profitable. The calves were roughed through the winter on the best available forage and with just enough supplementation to maintain normal growth. We reduced the number of cows we were running by about twenty percent to have forage for the yearlings for the additional seven months. Carrying our yearling cattle through the spring flush and selling them in the summer gave the effect of loading up on cattle (and forage demand) when forage was plentiful and reducing forage demand when forage was scarce. It also improved profit margins per calf, as the gains from weaning to feeder weight were cheaper than the gains from birth to weaning, which had to include the expense of maintaining the cow. The cattle breeding program was changed to produce a smaller, easier fleshing animal that could thrive on forage alone and winter more cheaply; hair sheep were added to the stocking mix to turn weeds into an asset and to provide another product not tied to the cattle cycle.

Animal health and nutrition were improved through closer attention to providing forage at the proper stage of growth and "on a clean plate" through good grazing management. The need for toxic pest control materials dropped, and general animal health improved dramatically after a program of planned high stock density grazing was established. Part of this was due to better hygiene and to leaving the pest organisms behind as the stock moved and part was due to forage becoming more diverse and being presented to the animals at the proper stage of growth. When we stopped the cattle spraying program, sand wasps, spiders, and other insect predators returned, and populations of horn flies, horseflies, and face flies dropped; we still had some of these pests, but not nearly as many as we'd had when we were spraying. Livestock maladies such as calf scours, retained placentas, hoof rot, and pink eye were greatly reduced through the combination of improved nutrition and clean pasture provided by planned grazing. At one time, I bought calf scour boluses by the gross, but in later years, it was an unusual event to find a calf with scours. Calving difficulties and the associated poor breed back and loss of both calves and cows were pretty well eliminated as we moved to smaller and earlier maturing cattle that were calving after the cows had access to twenty to thirty days of unlimited green forage. Major benefits continued to accrue as life in all of its diverse forms returned, and the effects of years of poisoning and tillage began to heal. Pests such as horn flies and internal parasites, which spend part of their life cycle in dung, were greatly reduced when exploding populations of dung beetles quickly buried the dung. All of the predator species, from scissortail fly catchers to sand wasps to carnivorous nematodes, returned to control their prey species, and the populations of pest species, from horseflies to heel flies to pecan case bearers, dropped and stabilized. The management that brought about these changes also greatly improved both the production and the stability of our pastures. All of this took time and many mistakes to accomplish, and it often felt as though we were taking one step forward and two backward, but in 2000, we realized that for all practical purposes, we had been organic for many years, and we made the few changes required to certify as organic. This allowed us to begin to produce and sell organic grass–finished beef and lamb at premium prices. What began as a cost-cutting program became a fascinating learning process that continues to this day. The program has been successful by all standards. Profitability increased dramatically, labor requirements were reduced, and, most importantly, our health as well as the health of our soils and our animals improved.

Factors That Contribute to the Profitability and Stability of Grazing Operations

CHAPTER 5

Get in Sync with Reality

All operations have a unique and limited set of resources with which to work; the quantity, quality, and cost of resources such as forage, capital, labor, equipment, and knowledge are different for each operation. The ranch that does the best job of matching enterprises or forms of production to the available resources is normally the most successful. Three types of factors (physical, financial, and human) should be considered in choosing enterprises to pursue and the amounts of resources to allocate to each. Trying to force a type of production that is not well adapted to the available resources in any one of these three areas is a recipe for failure. In the 1950s, I watched a ranch on the gulf coast of Texas travel annually to the production sales of reputation Angus breeders in the northeast and buy breeding stock. These fat, long-haired, and grain-fed cattle would literally melt away in the heat, humidity, parasites, and low fertility soils of the gulf coast, but the gulf coast ranchers kept buying more year after year. My current favorite example of this type of physical mismatch is the rather common practice of grazing goats on Bermuda grass pastures; pure stands of any grass are very poor choices for cattle or any other grazing animal, but for goats, they are a disaster. The goat is the most nutritionally demanding of all of the domestic animals; it is an extremely selective grazer that evolved eating the very nutritious and mineral-laden growing points of woody plants. Goats need browse and forbs to be healthy; they can live on grass alone but will not thrive.

Goats need brush; if you don't have brush, you don't need goats.

The animals used must be adapted to the climate and to the forage present on the ranch. Species of animals differ in diet preference: cattle, bison, horses, elk, and some antelope prefer grass with small amounts of forbs and browse; sheep prefer a mixture of forbs and grass with some browse; and goats and deer prefer browse and forbs with some grass. The same standards of quality that apply to grass hold true for forbs and browse; green and growing material is valuable, while old, senescent material is of little use to animals. Forcing animals to exist on an unnatural diet creates health problems for the animals and makes poor use of resources. Grain is an unnatural diet for all ruminants, as it acidifies the digestive tract, reducing the digestion of forage and creating health problems for both the animals and the people who eat the grain-fed meat. The current outbreak of E. coli bacteria from beef could be completely cured by feeding the animals more forage and less grain at least for the last several weeks before slaughter. Doing this would change the pH of the animals' digestive tracts from acid to basic or neutral and would kill the acid-loving E. coli that sickens people.

Over time, animals adapt to conditions as diverse as soil fertility, kinds and numbers of parasites, and climate. There is a steep learning curve and production lag time when animals are moved to forage radically different from that on which they were raised, as animals must learn what and how to graze in the new forage sward. This phenomenon is well illustrated by the work conducted at Utah State University, where cattle are taught to eat and thrive on weeds that are present in their grazing lands but not eaten. Proof that this is an educational process is seen as mothers pass the weed-eating trait on to their offspring and as other animals in the herd soon learn by observation to include such weeds in their diet. The wider the range of plants that is consumed by grazing animals, the more productive and stable the forage sward; this will be discussed more fully in the next chapter. Just as grazing animals must adjust to changes in diet, the rumen flora must make adjustments when new types of forage are ingested. Cattle people have long held the belief that cattle will do well if moved north and west but not if they are moved south and east. Moves from a highly mineralized soil to a depleted soil, from dry to wet, and from high elevation to low elevation are definitely hard on cattle. However, any move is disruptive to some extent, and the best policy is to work with cattle, sheep, or any other animals that are native and adapted to the local area and conditions.

You could roller skate through a buffalo herd, but going horseback is a heap better.

How well the types of production practiced fit the physical characteristics of the land and the climate will always be a prime factor in both the profitability and stability of an operation. For grazing operations, the critical factor is to match nutrient demand to nutrient production with forage. Production decisions involving livestock normally must be made well in advance of the growing season and require planning and knowledge of local conditions to arrive at the grazing season with the correct number of the right kind of animals. Questions such as the following must be asked: What is the normal forage production curve of the ranch, the quality curve of forage within a year, and the reliability of forage production both within a year and between years? Certainly, the quantity of forage grown is essential information, but equally important is when the forage is produced, the quality of the forage, and the length of the green season. The chart below would be representative of much of the southwest shortgrass country, but wide variations in timing and amount of precipitation are common. Know when you grow grass on your country!

GROWTH CURVE TYPICAL OF NATIVE WARM SEASON RANGE

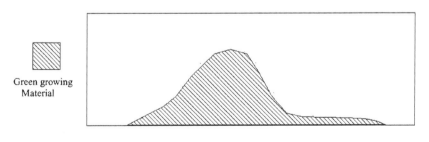

Reliability of forage production both within and between years must be considered when planning production programs. The higher the likelihood of substantial differences in year-to-year production of forage, the larger the percentage of the total stocking rate that should be made up of animals that can be disposed of quickly and without incurring a loss when forage production fails. The key to surviving a drought with as little damage as possible is early reduction in the demand for forage; the ways this can be done will be discussed in the chapter on planning for drought. The likelihood of drought varies widely by location, and a big part of developing a program suited to your conditions is knowing the probability of drought so that you can make reasonable plans.

Nobody wants an old cow with a small calf in the middle of a drought. A weaned calf or a yearling can get on a truck and go to where it is still raining.

The nutritional needs of the various classes of grazing stock must also fit the quality and timing of the forage produced. A native grass range in a twenty-inch rainfall belt where 75 percent of the year's growth takes place during ninety days in the spring would fit the needs of a spring calving cowherd much better than it would a grass-based dairy or grass-finished beef operation. The fit would be even better if a portion of the normal forage growth were devoted to a stocker enterprise that could be cut down or skipped in dry years.

If a significant part of the vegetation is made up of forbs or browse, adding sheep or goats to the stocking mix might increase profitability and stability while improving the land. By the same token, stocking sheep or goats on areas without forbs or browse increases costs, invites health and nutritional problems with the animals, and wastes resources. Sheep and goats offer some unique benefits; they consume more forbs and browse than cattle, and their market cycles are usually out of sync with the cattle cycle, which can offer a financial buffer during poor cattle prices. On the downside, costs of fencing and predation control are higher than for cattle, and the learning curve for the required husbandry can be steep for people with no sheep or goat experience. In the right situation with good management, sheep or goats can be more profitable than cattle while they improve the stability and production capacity of the land; in the wrong situation and management, they can be a financial and ecological disaster.

Which Is the Better Plan?

Production Plan A

High-quality purebred cattle producing premium-priced breeding stock Cows average 1300 lbs, and calves are born in late spring. Breeding bulls are sold as performance-tested three-year-olds, while breeding heifers are sold as pregnant long yearlings. Steers and cull heifers are sold retail as grass-finished beef. Cattle time- controlled grazed through twenty-four permanent paddocks per herd with temporary fencing used to increase paddock numbers when needed.

Production Plan B

Crossbred cows averaging 850 pounds calve in early fall. Calves are weaned in spring and sold. Cattle grazed through six paddocks.

See below for ranch description.

Ranch Description

Ranch is located in southwest Texas 80 miles from a town with seven thousand people. Area receives 12 inches of moisture in a normal year coming, mostly in July and August. Annual evaporation rate averages 12 feet. Stocking rate varies from AU/50 acres in good years to AU/200 acres in poor years. Stock water is from low-output wells pumped with windmills.
Which plan fits the ranch?

Periods when the nutritional requirements of animals are high, such as parturition and lactation, should be timed to occur when forage quantity and quality are normally high. The better the fit between the quantity, quality, timing of forage able to be harvested by grazing, and overall needs of the animals, the better the likelihood of a profitable and stable operation.

When a cow is grazing in the pasture, she is working for you. When you are feeding a cow, you are working for her.

Aside from its own intrinsic value, wildlife is increasingly important as an income-producing enterprise in many areas, and the needs of wildlife such as feed, habitat, and protection should also be addressed in the planning process. In most cases, this attention to the needs of wildlife will yield benefits not only to the wildlife but also to the operation as a whole, and for some operations, wildlife can produce as much or more net return than conventional livestock. The value for insect pest control of healthy populations of birds, for example, is seldom realized until the birds are gone. (To see a flock of half-grown wild turkey find a hatch of grasshoppers is to see true enthusiasm at work.) Even the despised coyote is beneficial as rabbit and rodent control as long as it is not killing stock. Most sheepherders don't believe it, but not all coyotes kill stock; if resident coyotes are not killing, they are the stockman's best friends, as they are highly territorial and will not allow strange coyotes that might be killers into their territory.

It is not possible for a country, a ranch, or a family to spend itself into prosperity.

Financial reality has a nasty habit of spoiling the fun at the most inopportune times; a good way to prevent this is to start with an honest assessment of the purpose of the ranch. If the ranch must be profitable and support a family, this is an entirely different situation than if the ranch is a long-term investment expected to appreciate over time or if it is a mechanism to provide a desired lifestyle. All three of these options can be valid given the right circumstances, but they have vastly different constraints and financial requirements. The critical difference is the percentage of the capital costs of the ranch that are to be paid out of profits from production. For quite a period in our history, agricultural land was valued primarily by its production potential; land that produced an annual crop worth one hundred dollars per acre would be priced at something close to one hundred dollars per acre. Land is still valued according to what it will produce, but the forms of production have changed. For most ranchland, the value of the forage produced has long since been surpassed by what the market is willing to pay for future monetary appreciation, enjoyment of wildlife, and lifestyle opportunities.

From the standpoint of logical financial management, the ownership of ranchland should be considered an enterprise totally separate from the operation of the ranch.

For the beginning rancher or any other rancher, the purchase of ranchland makes sense only if the cost of owning the land can be covered with income from other sources and if the rental value plus the value from appreciation, wildlife, or lifestyle makes an attractive and workable proposition. Although the market value of ranchland has increased greatly due to the other factors mentioned, the cost of grazing leases on ranchland in most areas is still heavily influenced by its production potential. It is usually possible to lease the grazing rights on a property for much less than the costs of owning the land. As more and more land is transferred to owners who are interested primarily in wildlife, investments, and lifestyle, the opportunities for good managers to acquire grazing land on favorable terms should increase; the operative word is good.

You can be a rancher without owning a ranch.
And
You can own a ranch without being a rancher.

Financial strength comes from making the best use of the resources available; the best use of resources will seldom be to use them to strive for maximum production. Money for profit-oriented operations is always limited, but even if it were not, it is only common sense to spend each dollar where it will do the most good financially at a particular point in time. This is a considerable change from the way many of us have spent money in the past. Too often, decisions on how production money is to be allocated have been made based on advice from outside parties anxious to sell something or on research-proven methods intended to increase production. If the goal is profit, the logical way to allocate money is to the areas that generate the most profit without degrading the resources.

It is critical that we know which parts of the operation make money and which parts lose money.

For financial purposes, it is necessary to separate the operation into enterprises; an enterprise being any endeavor that either generates or consumes money. If profitability is a goal, we have to know where income is generated and where it is spent. For example, a ranch might have a cow-calf enterprise, a replacement heifer enterprise, a hay-making enterprise, a stocker enterprise, a hunting enterprise, a wheat enterprise, and so on. These enterprises should be precisely defined; a registered cowherd is different from a commercial cowherd, and a stocker herd using home-raised calves is different from one using purchased calves. The purpose is to determine exactly which enterprises make money and which ones lose money, and the results of an enterprise analysis such as this can be surprising to some pretty good managers. The ranch will incur certain expenses regardless of the enterprises pursued, and these should be considered overhead and not prorated out to the various enterprises. Your CPA will scream at this, but the purpose of gross margin enterprise analysis is to determine what the various enterprises contribute to fixed expenses. These fixed expenses would be things such as a management draw, land payments, grass leases, taxes, insurance, hired labor with associated expense, and basic equipment such as pickups and trailers with their associated costs. These expenses are not parceled out to the various enterprises; however, any expense that would disappear if an enterprise were dropped is charged entirely to that enterprise. For example, hay equipment and the associated costs are charged to the hay enterprise; crop production equipment, its associated costs, and crop leases are charged to the crop. If an enterprise requires specialized or additional labor, these costs are charged to the enterprise as well. The objective is to understand which enterprises offer the greatest returns to the resources that make up the operation: the land, management, basic labor, and general-purpose equipment available to the ranch. Be realistic and hardnosed; if you make hay to feed cows, the cow enterprise should "buy" the hay from the hay enterprise at market value, and the hay enterprise must pay all costs associated with the specialized hay equipment and the fuel needed to run it. If you plant wheat for pasture and grain, the stocker enterprise should "pay" for the pasture just as it would "pay" the cow enterprise for calves. Spend some time breaking out enterprises, and be honest with yourself; do the cattle operations really need two heading horses, two heeling horses, and three calf-roping horses, or do you have another enterprise? It is possible—in fact common—to convert a ranch that is losing money into a profitable business by dropping one or two unprofitable enterprises. The most common culprits are making or feeding hay and registered anything's.

A guaranteed recipe for financial failure is to provide everything needed by animals that are not adapted to the situation.

The current rage over the "best management practices" theory illustrates why modern agriculture is in financial trouble; this theory claims that if every aspect of an agricultural operation is managed only by best management practices, profitability is bound to follow. In truth, this concept is a recipe for financial disaster, as many have found. It is not possible for a ranch, a family, or a country to spend its way into prosperity. For forty years, we have been deluged with advice designed to increase production. The assumption is that if production is high enough, profitability must follow. The flaw in this theory is that as production increases past a certain point, each new unit of production increases in cost due to the laws of diminishing returns and marginal reaction. Every farmer knows these laws are valid; as you increase any input, fertilizer, tillage, supplement, whatever, you get less response from each new unit of input and you increase the cost of production. Just because a practice will "cash flow" does not necessarily make it a good idea. At any point in time each dollar we have available has one use where if applied it will yield the greatest return. For example, a dollar spent on weed spray is not available for fencing or grazing management costs, even though these could yield higher returns in several ways and, with good management, could be a permanent solution to the "weed problem."

The amount of money coming in is not nearly as important as the difference between what is coming in and what is going out.

I am not picking on the weed spray salesman; he has a product to sell and is certainly going to present it in the best possible light. This is great; it is an example of the free market capitalism that has given us the most robust economy in the world. However, taking advice on what to buy from the people trying to sell you things might not always be the best idea.

All agriculture is inherently risky, as it is subject to factors such as weather, markets, and politics, over which the producer has little influence. When profit margins are thin is a poor time to increase inputs in the hope of increasing production. Increased inputs at this stage are valid only when they decrease the cost per unit of production or reduce the likelihood of production failure. Profitability comes about not so much from high production as from a wide margin between income and expense.

I have never met a salesman who thought that I did not need his product.

When times are tough, managers' first thoughts should be how to cut costs without cutting income. Purchased inputs are a good place to start; much of the money spent on inputs can be saved by making changes in management. Supplemental feed costs can be reduced dramatically when a program of planned time controlled grazing is implemented and weed control is better achieved with grazing management than with poisons. A large portion of the money spent on pest and disease control is wasted as it is directed at treating symptoms rather than at addressing the root problems. Treatments designed to kill the pests are seldom effective in the long term and often the effects of the treatment are worse than the malady. Any practice that destroys life should be carefully monitored to determine whether its long-term, overall effects are the ones desired. Problems with pest organisms are symptoms of poor management and can often be offset most effectively by changes in management. There are no simple "silver-bullet" solutions to the problems of agriculture. Agriculture is a biological process subject to all of the effects of weather, competition for resources, and other natural phenomena while also being impacted by economic and social factors. Most production problems are merely symptoms of underlying flaws in management. Calves dying of pneumonia in the snow can be treated with nutrition drenches, special warming blankets, and calving barns—but wouldn't it be better to just calve in warm weather? Problems with pest organisms such as weeds, brush, insects, and disease microbes usually occur because our management has simplified the local environment and disturbed the system of checks and balances that nature uses to control pests. Weeds become excessive not because weed seed is present but rather because growing conditions favor their abilities more than they favor the abilities of the plants we are trying to grow. If what is growing on the land is not what is desired, change the long-term growing conditions to favor what is desired..

A second, closely related reason for pest explosion is stress-weakened immune systems caused by flaws in management. The example that comes to mind is shipping fever, or BRD. We pull calves off their mothers and stick them, bawling and scared to death, in a dusty lot where nothing, even the feed and water, is familiar; then we wonder why they get sick. We could give them inoculations weeks ahead of weaning time and start them eating specially medicated receiving rations, or we could simply wean them on pasture across a fence from their mothers, where everything is familiar and there is no reason to be afraid. If they are not afraid, they are not stressed and if they are not stressed, they won't get sick.

A major fault in the way technology is applied is that we tend to use it as an excuse to do what is comfortable rather than what we think is right; if we do what everyone else is doing and fail, there is no shame involved. If we do things radically differently from our neighbors, we will be pointed out as odd even when we succeed. Thus, if we have insect damage, we don't reason why; we just spray. If we are short on grass, it is easier to call the fertilizer man than to figure out why we are short.

Planning is the most important work that a manager can do. Planning production is fun, but planning profitability is more rewarding. Pay yourself first; plan profit into your program just as you plan to pay the mortgage. If you cannot realistically plan in a profit, scrap the whole plan, and start over; planning a profit on paper does not guarantee success, but if you cannot make the plan work on paper, failure is almost guaranteed. Most crises occur because of a lack of planning, and most of the rest occur because of a failure to monitor and correct the plan as it proceeds. If you are a good problem solver, you are a poor manager. Good managers plan to prevent problems and don't get enough practice at problem solving to get good at it.

Planning is the most important work a manager can do.

On occasion, I have been accused of being a Luddite—one of the bunch that went around wrecking weaving machinery in England at the start of the industrial revolution because they didn't want technology to replace their jobs. I don't think this is a fair criticism. I am not opposed to all technology; I am opposed only to technology that has costs that outweigh its benefits. Sometimes the real cost of a tool is hard to see, because it is difficult to determine its long-term, overall effects. As an example, when we put a hundred pounds of nitrogen fertilizer per acre on pasture, we can see results almost immediately. The grass greens up and starts growing rapidly, and it is an easy process to measure how much more grass and how much more protein we have produced by using the fertilizer. We know what we paid for the fertilizer, and we can figure out what the increased feed is worth, so it should be simple to decide whether we should use the fertilizer. It would be an easy decision if this was the whole picture, but every decision we make has effects beyond the immediately apparent financial situation. Let's stay with the financial aspect of this example for a moment; the value of forage depends on the season of its growth, and winter forage is normally more valuable than summer forage due to supply and demand. Forage also has a higher value if it can be harvested by grazing, as no mechanical method of harvest can compete in cost with the grazing animal. Therefore, our decision is a little more complicated; is fertilizing the way to make more forage available for grazing at a lower cost during the lean season? If we put nitrogen on grass in the spring, we increase the amount of forage during a period when we normally have a forage surplus. Have we helped or made the situation worse by increasing the surplus? Our normal reaction to this situation is to make hay of the surplus. Good decision or bad? That depends; could we have made enough hay from the natural spring flush or perhaps by fertilizing half as much area? Could we decrease the amount of hay needed by using better grazing management or by using the money spent on nitrogen to seed cool season legumes? Would we be ahead financially if we reduce stocking rate to provide more forage per animal, or could we plan to change the time of heaviest demand for forage to coincide with the time that forage is most abundant?

From a biological standpoint, adding nitrogen fertilizer has several effects beyond the obvious ones. I once heard Dr. Bill Knight of Mississippi State say that sixteen pounds of actual nitrogen would kill every rhizobia bacterium on an acre of sandy soil. Heavy or prolonged use of nitrogen fertilizer reduces soil organic matter content and soil life. It also has an acidifying effect on the soil, which can cause some minerals to become less available while others become more available. We could go on and on listing pros and cons of nitrogen fertilizer, and we could and should do this with any technology that we are considering using. The effects of technology on a biological system are never simple; they will always have far-reaching side effects that are difficult to determine. The manager who does the best job will be the one who can best evaluate the long-term overall effects of practices and use those that best meet the needs and goals of the operation. Sometimes this means having to rethink things that we have been doing for a long time; this is never a comfortable process, and it is even worse if it entails things that "everybody" does.

This brings us to my biggest complaint about technology as we use it; grazing, or any aspect of agriculture, is a complex business, and too often we use technology as a crutch to let us limp around problems rather than thinking through the complexities and solving the problems.

It is our nature to take direct action: kill that weed, plow that field, do something! It is very tempting to believe that somewhere out there is a silver bullet that will solve all of our weed problems, parasite problems, or soil compaction problems forever. Reality is that silver bullets work only for vampires. Or is it werewolves?

A third factor that plays a large role in the success or failure of any operation is the strengths and weaknesses and the likes and dislikes of the people involved. The availability and quality of both management and labor are huge factors that must be considered when planning an operation. Exceptionable knowledge and commitment of the people involved is usually most needed and best rewarded in intensive operations, but even the simplest of commodity-based ranches requires an above-average level of knowledge to compete in today's low-margin world. It is a fact of life that most ranching operations are not profitable.

Managing according to the conventional wisdom, by which failing enterprises are managed, is not likely to produce success.

Alan Nation coined the term "knowledge-based ranching" in his book of the same name, and he is correct in his belief that knowledge of the entire ranching process and the ability to apply that knowledge is the key to success. Profitability and stability are much more likely to occur when both management and labor have a good understanding of the ecological, financial, and human aspects of ranching. Management occurs on three levels; strategy (goals), tactics (planning), and implementation (practices and techniques); to be valid, a decision on any of these levels must be logical from the standpoints of all three considerations—ecological, financial, and sociological. It is sad but true that in modern agriculture, many decisions have been made and are still being made based solely on short-term financial considerations. True profitability, profitability stable over the long term, can be achieved only when the primary focus of management is the well-being of all aspects of the operation.

There are opportunities in niche markets, such as grass-finished beef, grass-based dairying, and organic meat production, but these endeavors require knowledge and skills in many different areas; in addition, they have rigorous forage requirements and normally need to be located close to population centers. The success or failure of such an endeavor will depend largely on how well the requirements are understood and whether the required knowledge and skills are available. The time to acquire the necessary knowledge and skills is before putting money on the line. The availability and cost of capital and the availability and quality of labor will always be major factors in determining the viability of enterprises.

It is usually a lot easier to get a loan than it is to pay it back.

It is important that everyone involved believe in the program; not everyone must have all of the required knowledge and skills, but everyone must believe that the program is feasible and that the details are important. It is not uncommon to see a well-thought-out grazing program fail because people on the ground who are doing the day-to-day work don't like or believe in what they are supposed to be doing.

Perhaps just as important as knowledge and skills are the likes and dislikes of the people involved. These personal goals, the likes and dislikes of the operators, should play a large role in the process of choosing a mix of enterprises that fits the reality of the situation. Few people will make a success of a venture that they do not enjoy or that does not fulfill their personal desires. Life is too short to spend doing something that you don't like. This is particularly true when working with animals; people who do not like animals should not work with animals, as it will only make both of them miserable. I read a study recently from England that said dairy cows that were named and addressed by their names gave more milk and were healthier than those without names.

I have never been much of a rough stock rider, but there was a time when I prided myself on being able to ride and get work out of horses that other people could not ride. I was able to do this not because of riding skill but because I usually could convince a horse that I liked him and was not a threat to his well-being. In the interest of honesty, I cultivated this skill to a higher degree as I got older and the ground got harder.

The problems that prevent success in endeavors are usually people problems. Very few issues can retard progress and take enjoyment out of an operation like having one or more people around who constantly cause conflict and division. I was close to middle age before I finally learned that some people are simply not worth what they cost in terms of headaches and heartburn. As the old timers said, "Ride for the brand or ride on down the road." Some employee issues can be fixed through training and education, but dishonesty, laziness, and bad attitude are usually not fixable. I would ten-to-one prefer to have an employee who doesn't know much but is willing and wants to learn than one who knows it all but doesn't care.

One of the toughest problems, and one that is very common in multigenerational operations, is the "young bull-old bull" syndrome, in which a younger generation is coming up in the business while the older generation is still active. Regardless of how much love and respect parents have for their offspring, and vice versa, the differences in age and experiences will cause conflicts. I don't claim to know how to solve these kinds of conflicts but there are people out there who do. The first step to take, if such a problem exists, is to discuss the situation together; you may find that the other party was unaware that a problem existed and is willing to work together to fix it without outside intervention.

CHAPTER 6

Maximize Biological Diversity

Biodiversity means simply that the life in an area consists of a large number of different types of life forms, with plants and animals of all sizes represented by many different kinds of organisms. The term has a bad connotation for many farmers and ranchers, but it should not. Biodiversity is what gives stability and long-term productivity to a local environment.

Every kind of organism in the world has unique needs and the abilities to meet those needs. Organisms are not "good" and "bad"; they are only different.

Radical environmentalists (who, in my opinion, are frequently motivated more by political zeal than by any real search for truth) have made biodiversity their war cry in their attempts to destroy the capitalist system. Whether the stated goal is to halt logging in the northwest, remove livestock from public land, or force a reduction in the use of fossil fuel, their only interest in science is to find some "useful idiot" academic type who will help them demonize their opposition. An example is the assertion that the science of manmade global warming is settled because a consensus has been reached by a bunch of scientists; science is settled by discovering facts, not by reaching a consensus. It was not that long ago that almost all scientists agreed that the Earth was flat. Most of those who did not agree kept their mouths shut, as open opposition to the powers that be was a good way to get burned at the stake; mostly today, nonbelievers just get their academic careers destroyed.

Most of the radicals' chosen arguments rely on emotion rather than logic and promise immediate and total doom if their position is not wholly accepted and acted on immediately. Theirs is a perverted view of how the natural world operates, and it is quite obvious that they do not understand the benefits of individual liberty and limited government enshrined in the United States Constitution. True husbandry of the environment is much more likely to occur with a free people protecting their own well-being and that of their descendants than at the hands of a politically motivated bureaucracy.

Having vented my gall on the radical environmentalists, I must in fairness say that too often we farmers and ranchers have taken an equally short-sighted view of the natural world. We tend to make value judgments based upon the short term rather than on the long term: grass is good—weeds are bad; livestock is good—varmints are bad. Insects, bacteria, predators of all kinds, and any plant, animal or microbe that doesn't have an immediately evident value are apt to be put in the villain category. I was speaking to a group of farmers and ranchers in northern Missouri several years ago, and I spent some time discussing the benefits of high biodiversity to agriculture; after my talk, a group of people were standing around me asking questions and making comments when a very agitated man got right in my face and shouted, "What the hell good is a butterfly?" It turned out that he planted corn genetically modified to contain the insect killer Bacillus thuringiensis (Bt) and he was highly incensed about reports that the pollen from Bt corn might be killing butterflies. He felt that by pointing out the value of biodiversity, I was threatening his right to use what poisons he pleased. I started to explain the tremendous benefits that butterflies and other moth-like creatures bring to plant pollination, but he was not interested and stormed off.

Long-term productivity and stability result when all available resources are fully utilized without any being overutilized.

Biodiversity gives complexity to the populations of plants, animals, and microbes that make up the life present in an area. This complexity brings stability and long-term productivity to an area through full utilization of all of the resources available without overutilization of any the resources and through the formation of a web of mutually beneficial relationships between the various life forms. By its life and death, every organism from microbe to large animal will contribute to the improvement of the local environment; this improvement will occur provided that we do not short circuit the process through poor management. The buildup over time of all of the tiny bits of improvement added by each of the innumerable organisms increases biological capital.

The improvement from biodiversity will take place as advances in the health of the four ecological processes: water cycle, nutrient cycle, energy flow, and biological succession.

Biological capital is biodiversity plus the long-term effects of having biodiversity; it is soil with high organic content that has excellent tilth and structure and holds a lot of its mineral content in organic form, and it is diverse and healthy populations of plants and animals both in and on the soil made up of healthy individuals. Biological capital is what allows the ecological processes (water cycle, nutrient cycle, and energy flow) to function properly, and it provides a system of natural checks and balances that limits the populations of pest organisms.

Complexity builds productivity and stability.

Biological capital is wealth in the truest form and is vital not only to agriculture but to society as a whole. When biodiversity is high throughout the entire soil-plant-animal complex, both productivity and stability will be high. Weeds, disease, parasites, and pest organisms of all types will still be present but not in concentrations high enough to interfere with the functions of the local environment or of the humans living in the environment.

Biological capital is real wealth and is at least as valuable as fiscal capital.

I saw an excellent example of the benefits of biological capital in 1990 when we had record flooding in the Red River Valley of Oklahoma and Texas. About 85 percent of our ranch went under water in early May, and some of this water stayed for three weeks. When the water finally did recede, we saw a dramatic difference in land response depending on previous treatment. On areas that had recently been in cultivation and had low soil organic content, nothing grew except a few annual grasses and weeds, and these did not sprout until late August. The soil was compacted and had an odor similar to that of sewerage; cattle put in these areas after the sparse forage appeared would not graze; they would smell the ground and bawl and pace the fences until they were removed. In areas where there had been mixtures of perennial and annual plants and that had been under high stock density for several years, as the water receded, earthworms opened their burrows at the water's edge. All of the plants present when the flood came were killed by long-term submersion, but new plants from seed and perhaps from rhizomes (I failed to check this) quickly established, and we had quality forage in about sixty days. Some of the recently cultivated land that was most severely damaged was under water for only seven or eight days, while some of the land that was in good condition due to a history of high stock density grazing was under water for twenty-one days but recovered much more quickly.

The health and productivity of an environment is determined by the conditions of the ecological processes: water cycle, mineral cycle, energy flow, and biological succession.

The water cycle will become more effective when soil organic content increases through the life processes of and the death, and decay of living organisms both in and on the soil; this increase in organic matter improves the ability of the soil to take in and retain both water and air. This increased capacity occurs through improved soil structure due to organic compounds that hold soil particles together in granules with open pore space between the granules; it also comes from the ability of soil organic matter to absorb many times its own weight in water. The cycle is made more effective when the soil surface is kept covered with organic material either alive or dead. A covered surface loses less water to the evaporation that occurs due to high temperatures or due to wind effects and it is more effective in water infiltration. Increased soil organic content means more soil life, from bacteria to earthworms, and the increased amount of soil life brings further improvement in both the water cycle and the nutrient cycle. In some semiarid and arid regions with a lot of bare ground, only 20–30 percent of the precipitation is ever taken up by plants; a large portion runs off without ever soaking into the ground or evaporates out of the ground before being captured by plants. Bare ground causes plant use of water to be inefficient due to high soil temperatures. I remember being a barefoot kid with feet as tough as rawhide but on bare ground, I still had to run from one patch of grass or shade to the next to keep from burning my feet. We cannot control the amount of precipitation, but by improving the water cycle, it is possible to double or triple the amount of water available to grow grass. Grazing management is one of the most effective and least costly means of improving the water cycle of an area. Planned grazing properly designed, implemented, and monitored can greatly reduce the amount of water that runs off an area and the amount that is lost to evaporation. The water retained on the land is available to grow vegetation and to recharge underground aquifers and does not contribute to flooding or soil erosion. It is common for ranches to have long- dry springs begin flowing again and to see marked reduction in dryness-adapted plants, such as cactus and bull nettles, when their water cycle improves through better grazing management.

Tend the health of your water cycle, and double or triple the amount of water available to grow grass.

Mineral nutrients cycle through the soil-plant-animal complex. They are taken up by plants and formed into compounds; eaten as forage by grazers large and small; excreted in manure and urine; broken down by insects, bacteria, and fungi; and finally, taken up again by plants. The amount and types of minerals in the cycle are dependent upon the mineral content of the local soils, but the health of the local soil life, the diversity of the vegetation sward, and how uniformly the vegetation is utilized by grazing animals are also critical factors. The greater the amounts of mineral nutrients cycling in a system, the more productive that system. Minerals tied up in old, senescent forage, ungrazed browse or forbs, or dead plant material that decomposes chemically rather than biologically are not available to produce growth. Tremendous growth is possible with rather limited amounts of minerals if the mineral cycle is functioning well; the vast amounts of vegetation in rainforests grow not because the soil is fertile but because the minerals that are present are in constant cycle within the system. In a rain forest, when a leaf is eaten or a tree branch falls, the minerals contained in the material are immediately ingested by living organisms of some sort and kept in the cycle. The mineral cycle of healthy, well-managed grassland can function in an even more efficient and effective manner, because minerals are not locked away in long-living plants. At least 90 percent of all minerals eaten by a cow are excreted in the manure and urine; for most minerals, the percentage will be even higher. If the manure and urine fall on healthy soil covered with organic matter, the nutrients in this material will very rapidly be taken in by live organisms of some sort and thus will stay available to the system. If the manure and urine fall on bare soil with little soil life, most of the nutrients will be lost to chemical decomposition or will be carried away by wind or water.

Life - Growth - Death - Decay - Life
It must be an unbroken cycle.

Most of the need for fertilizer materials and livestock mineral supplements stems from an ineffective mineral cycle in which nutrients are either lost to the system due to poor soil management or sold off the land in products. When we sell 1000 pounds of beef, we sell 820 pounds of water, about 9 pounds of calcium, about 5 pounds of phosphorus, about 4 pounds of potassium, and lower amounts of other minerals such as magnesium and zinc. We also sell large amounts of carbon, which we get free from the air, and nitrogen, which we can also get free from the air (with the help of legumes, rhizobia bacteria, and other natural fixers of free nitrogen, such as blue-green algae and azotobacter bacteria). These materials are free and inexhaustible, provided we don't short circuit the system with poisons or poor management and kill off our helpers. If we were to sell the amount of grass needed to grow 1000 pounds of beef as hay, we would sell about 44 pounds of calcium, about 18 pounds of phosphorus, about 158 pounds of potassium, about 13 pounds of magnesium, and corresponding amounts of other minerals. Grazing ruminant animals and selling the meat, milk, and fiber products places a low demand on the mineral content of our soils, and these minerals can normally be replaced through good husbandry of our soil life and very low mineral supplementation. The Amish and some other good farmers figured this out long ago; they will sell you meat, milk, and eggs, but not hay or grain. One of the most glaring flaws of modern agriculture, and one that will eventually prove fatal, occurred when animals were removed from the production system. Removing animals from the land and concentrating them at industrial sites changed them from being vital links in the maintenance of soil fertility and productivity to being agents of pollution. Production can be achieved with enough technology and enough purchased inputs, but the economic and ecological costs of industrialized agriculture, without animals to recycle nutrients and decompose plant material, will eventually prove too high to sustain.

The mineral nutrition available to plants is more dependent on the health of the mineral cycle than on the quantity of minerals in the soil.

All energy that we use, with the possible exception of atomic energy, is derived from the sun as solar energy. Even petroleum and coal are simply fossilized solar energy that fell on the earth long ago. Unlike water and mineral nutrients, energy does not cycle but rather flows through the system. In a normal scenario, solar energy as sunlight is converted into biological energy by green plants through photosynthesis; this green forage is consumed by grazing animals, and the energy is used to produce animal tissue. The animal tissue will later be consumed by other organisms, from man to bacteria, with the products of this consumption finally broken down by a series of soil organisms. At each stage, when the energy containing material changes form, some energy is transferred to the atmosphere as heat; energy is not destroyed in the process but rather changes form and is eventually lost to the system. This flow through with energy being lost to the system at each stage means that the system requires constant infusions of energy to remain healthy. The higher the amount of solar energy captured and converted to biological energy, the higher the production of life by the system. The higher the production of energy, the more energy we can harvest for our own use without degrading the soil-plant-animal complex.

As graziers, our primary business is the collection of solar energy and the conversion of this energy into biological energy—and finally into wealth.

Increasing the amount of energy flow within a system is straightforward; the key is to have green and growing vegetation present for as much of the time when photosynthesis is possible as is feasible. Sunlight is converted to biological energy only when it falls on green material, and the conversion is most effective when the green material is still actively growing; as we find when we get older, young tissue is more effective than old tissue in a lot of ways. Sunlight that falls on bare ground or on senescent plant material is wasted as far as increasing energy flow. A common means by which energy flow to an area is reduced is abusive grazing. Whenever the area of photosynthetically functional leaf is reduced, the amount of energy converted is reduced. Continuous grazing will reduce energy conversion by two methods: (1) plants being overgrazed and thus losing leaf area and (2) plants being underutilized, which means that much of their leaf area will become senescent and inefficient at photosynthesis.

Whenever the area of photosynthetically functional leaf is reduced, the amount of energy converted is reduced.

A major management goal should be to have green and growing plant material present at all times when photosynthesis is possible. In addition, vegetation swards containing a mixture of plants with different growth habits (warm season, cool season, grasses, forbs both leguminous and nonleguminous, and browse plants) normally capture and convert more solar energy than vegetation swards with fewer types of plants. Adopting the concept promoted by chemical companies (that the best pastures are composed of only "clean stands" of grass) damages ranches in a number of ways both financially and ecologically. Biologically diverse forage swards are more productive, more stable, and far less costly to grow than grass-only swards.

I still use the term biological succession, even though some have replaced it in the list of biological processes with community dynamics. Some people feel that the term succession might suggest a march toward some ultimate "climax" condition. To me, the term biological succession means simply a process by which those organisms that are best adapted to a given set of environmental conditions will, over time, prevail. I have no real quarrel with terminology, although "community dynamics" does sound a little like a reality show for community organizers. I am cool as long as we are talking about the concept of life within an area changing to adjust to changes in the long-term growing conditions. Changes in long-term growing conditions are brought about by changes in the water cycle, mineral cycle, and energy flow of an area. Changes occur when poorly adapted organisms are replaced by others better suited to the new environment, but changes can also occur through change in individual organisms. An example would be the morphing of blue grama from an upright, mid-height grass to a low-growing, sod type in response to prolonged continuous grazing—and its return to mid-grass status when grazing management is improved. In this example, continuous grazing would bring about poorer long-term growing conditions by reducing the amount of energy flow through the system and by reducing growth, lowering the amount of minerals in cycle. There is no best, or climax, state of succession for an area but there are states that will tend to come about due to conditions prevalent in an area. Much of the eastern United States has climatic and soil conditions that favor the formation of hardwood forests; these areas have rainfall patterns that are fairly uniform both between and within years and thus favor vegetation made up of long-lived individual plants. Farther west, precipitation patterns become more erratic, and temperatures vary more widely; these conditions favor plant communities consisting of shorter-lived plants that can reach reproductive maturity more quickly. Perhaps the most valuable knowledge a rancher could have would be to understand how management techniques can be used to promote good, long-term growing conditions through the ecological processes. An extremely valuable source of this knowledge is *Holistic Management* by Savory and Butterfield.

Good, bad, or indifferent, your management will determine the condition of your country in the future.

If the goal includes healthy animals, healthy land, and healthy finances, the most successful ranch over the long term will be the one in which the health of all parts of the local ecological system is carefully guarded and biological capital is valued at least as highly as is financial capital. Increasing biodiversity is the only means by which ranching, or any agricultural endeavor, can develop the triple bottom line of being ecologically and financially sound while at the same time meeting the needs of the people and the animals involved.

There are no junk organisms or invasive species, only organisms adapted to a certain set of environmental conditions. If you don't like what is present, change the conditions.

Conventional agricultural practice is, almost universally, destructive to biodiversity and to biological capital. It is tailored toward the production of monocultures of various plants and to the concentration of large numbers of animals in small areas in search of efficiency of production. Reliance on monocultures and common production practices such as pesticide use and tillage greatly reduce the number of kinds of plants, animals, and microorganisms present on the land and in the soil. This lack of biological diversity in the production units sets the scene for many of the problems encountered in modern agriculture, as simple communities of plants and animals are inherently unstable. The constant onslaught of weeds, diseases, and pest animals that plague the modern farm or ranch are nothing more than nature's efforts to put in place organisms capable of surviving in the existing conditions and, by doing so, to reestablish a functioning community of plants and animals that uses and reuses all of the available resources of water, energy, minerals, and space while wasting none of them. Encouraging biodiversity is the only way to promote agricultural production that is high yield and both financially and ecologically stable. This does not mean that we can no longer have monocultures of corn, wheat, and the like, but it does mean that we must recognize how the practices used to produce these monocultures impact the local environment and learn to offset the negative effects with management. Argentina has a very well-developed system of agriculture that relies on pasture being rotated into crop production and back into pasture; soil health and productivity, weed control, and profitability are all achieved without heavy input usage. This type of program was once common in North America, and I believe it should be again.

The key to stable productivity is full utilization of all available resources without overutilization of any resource

There are no monocultures in nature, because no single species of life can utilize all of the resources of water, sunlight, space, and mineral nutrients that are available in an area over the course of time. Every type of organism has requirements and the abilities to fulfill these requirements that are different from those of even its close kin. Nature abhors wasted resources and, left alone, will fill all empty ecological niches with life that is adapted to that particular set of environmental conditions. These special adaptations include the obvious traits such as the ability of legumes and rhizobia bacteria to symbiotically fix nitrogen or the adaptation of some plants to warm-season growth and others to cool-season growth. Many less-apparent traits exist as well, such as the ability of some forbs to concentrate certain minerals that are in short supply, the ability of tap-rooted plants to break hard pans and bring unused deep soil profiles into the system, and the ability of mycorrhizal fungi to multiply the effective root system of plants. Healthy soil and healthy plant communities are formed by the minute contributions of innumerable organisms living, dying, and decaying over time.

Every living organism has a unique set of needs and the abilities required to meet these needs.

All plants have, in larger or smaller quantities, antiquality factors or toxins that act as intake inhibitors for grazing animals, from worms to elephants; many of these compounds have evolved in diverse combinations in different plants. When animals consume too much of them, these compounds cause a feedback reaction that says, "Whoa, remember how you felt the last time you ate a bunch of that stuff." Fred Provenza and his crew at Utah State have found that palatability for grazing animals is less about taste (sweet, sour, bitter, etc.) and more about the aftereffects of having consumed a plant. If the consumption of a plant causes distress, that plant will be deemed "unpalatable" and will be consumed in smaller quantities. Plants high in needed nutrients give better aftereffects than do plants with lower levels of nutrients and thus are perceived as "more palatable" by the animals. One of the outgrowths of this research is the discovery that some of these antiquality compounds (toxins) counteract the effects of other antiquality compounds. Animals that are able to select from a large number of different kinds of plants can eat and utilize more high-toxin plants, because the toxins in one kind of plant offset the effects of toxins from other types of plants. The amount of sage brush, sericia lespedeza or other high toxin plants that an animal will utilize and benefit from is greatly increased when a variety of other feeds are available. If the variety of plants in the pasture sward is reduced by poor grazing management, these high toxin plants increase to pest concentrations because animals are unable to utilize them without other plants available to offset their toxins. Using time controlled grazing to keep nutrient content high, will also allow animals to take in more of the plants containing antiquality factors without distress.

Having a wide diversity of plants utilized by grazing animals will increase forage intake and animal production while increasing forage production by bringing more types of plants into the grazing sward. One of the less-recognized benefits of time-controlled high stock density grazing is that it encourages animals to broaden the range of plants that they habitually graze. This behavior can be taught to grazing animals, and they, in turn, will teach it to their offspring and to other members of the herd. Dr. Bob Steger told me years ago that animals grazing at high stock density would utilize some mesquite in their diets; I knew this was wrong, because, "nothing will eat that stuff." I had to eat crow and admit that he was right when I saw the evidence with my own eyes. The amount of a high toxin plant that animals will eat depends in part on the variety of plants on offer. I have a picture of a mesquite tree that shows the classic "overgrazed plant shape" where every growing point has been cropped back repeatedly until the plant resembles a ball. This tree is one of a very few mesquites on a twelve-thousand-acre ranch, and everything that comes by takes a bite, even if only out of curiosity. I like jalapeño peppers but I would not like them as much if they made up half my diet.

When more of the various types of plants in an area are grazed, the carrying capacity of that area goes up, and its health increases. When a large percentage of the plants in an area are grazed, more mineral nutrients will be cycled through the system instead of being locked away in senescent plant material, and more solar energy will be captured, because the regrowth of grazed plants is more effective than old, ungrazed material in converting solar energy to biological energy.

Life begets life
The more types of life present in an area, the smaller the amount of resources wasted or unavailable to the system.

There are organisms capable of surviving under any set of environmental circumstances found on Earth, and, simply by existing, each of these organisms improves conditions for the whole soil-plant-animal complex. There are no junk organisms; there are only organisms able to function in a particular set of circumstances.

> *Large numbers of the plants we label weeds, the microorganisms we label diseases, and the animals we label pests are not of themselves problems; rather, they are symptoms of deeper problems in the health of our local environment.*

In many (perhaps most) cases, these problems arise from a lack of complexity in the local system. In highly complex systems, one species of organisms seldom reaches population densities high enough to be considered a pest due to competition from other species.

> *Complexity builds stability; all simple communities, whether plant or animal, are unstable and subject to wide swings in population numbers.*

Plants and animals that are noxious in high concentrations are valuable to the system when present in lower concentrations. In a pasture sward that is 70 percent ragweed, the ragweed is properly named, but if the sward contains only .5 percent, the ragweed becomes a valuable member of the community, providing soil improvement, forage, and feed for birds and other small animals. Grasshoppers are excellent sources of protein for a wide range of animals and increase the rate of nutrients cycling in the system by digesting fibrous plant material; they become pests only when conditions such as low soil biological activity allow their population to explode. Weeds become plentiful not because weed seed is present but because the conditions of the vegetative sward and the conditions of the soil fit the special abilities of the weeds better than they fit the abilities of more desirable plants. Examples of such conditions include bare ground, low soil organic matter content, mineral nutrients lacking or tied up, compacted soil layers, lack of plant diversity, and continuous exposure to defoliation. The conditions that favor the explosion of pest species are, with some rare exceptions, brought about by how the land is managed and can be corrected by improved management. The hallmark of the good manager is the ability to understand the overall long-term effects of management practices. In most cases, understanding is gained by applying a practice and then monitoring the results over time, but with thought, it is possible to predict the effects of practices without having to actually apply them. The Holistic Management Model, as taught by Alan Savory in *Holistic Management*, is invaluable for this purpose.

The good manager is not a good problem solver; the good manager prevents problems and doesn't get enough practice to become good at solving them.

If the goal is a productive and stable landscape, then any practice that reduces the effectiveness of the water cycle, reduces the quantity or rate of cycling of nutrients being cycled, or reduces the amount of energy flowing through the system is likely not a good choice. Simplifying the system by reducing the number of species present brings about all of these deleterious effects in the long term. Conventional agriculture stresses increasing production and short-term financial returns (cash flow) as decision-making criteria. Although these are important considerations, they do not rate overriding importance when dealing with a complex enterprise such as agriculture, which has financial, ecological, and sociological components. For a decision to be valid, it must be beneficial to all three of these components. Many common agricultural practices offer short-term benefits that must be paid for with long-term costs; this is particularly true of practices aimed at killing pest organisms such as weeds, insects, and predators. If the practice is effective in reducing the numbers of the targeted species, it is certain to reduce the complexity of the system by reducing the number of species present, thus setting the scene for even worse outbreaks of pest organisms.

Swapping long-term costs for short-term benefits is always a poor trade.

In plant communities, there is strong competition among plants for moisture, sunlight, mineral nutrients, and favorable germination sites; in addition, there is growth limiting pressure from diseases and plant-eating animals. In a plant community simplified by herbicides, tillage, or other practices, many of the plants present are similar and thus compete for resources, using the same or a very similar set of abilities; thus, some resources are overtaxed, while others go unused. The unused resources are wasted with regard to the conversion of solar energy to biological energy, so the amount of energy available to the entire system is reduced as complexity decreases. As the number of species is reduced, the number of individuals within a species population increases. The concentration of large numbers of the same type organism allows diseases, parasites, and predators of the species to flourish, which creates an unstable situation in which the species become prone to wide swings in population numbers and periodic, catastrophic population die-offs.

CHAPTER 7

Planned Grazing

Planned grazing management is the most powerful and cost-effective tool available for increasing both the profitability and stability of ranching operations. The unique value of grazing management is that, when properly applied and monitored, it can simultaneously increase financial profitability and ecological health. You can literally have your cake and eat it too! That is a strong statement, but it is absolutely true. The mechanics of how a program is designed and operated vary widely depending on the area and climate, but the principles are universally valid. I cannot imagine trying to make a profit with a grazing operation in any area without a time-controlled, planned grazing program. Equally daunting would be managing an area to increase the health of a watershed or to improve wildlife habitat without planned grazing. Planned grazing is not just another term for what is commonly called rotational grazing, and is it not one of the many grazing systems. Planned grazing is fundamentally different from any of these because with planned grazing, the focus is always on what is to be accomplished. We plan for what we want to happen and follow up to make certain that it does happen; the process is under constant review, and changes are made in the plan as soon as faults are recognized. There is nothing sacred about the process; the process is only a means to accomplish predetermined desired results. As ranching is an ongoing operation that takes place in a constantly changing environment, management must be able to adjust rapidly to both normal and unusual changes in ecological and financial factors. At any point in time you will be managing to see that several different things are being accomplished simultaneously and the emphasis on what is most important at that time will change over time and as conditions change.

With planned grazing, the focus is always on what is to be accomplished.

Example: Today you need to promote the highest feasible gain on the yearling stocker cattle while simultaneously increasing the health and vigor of the forage base; when the yearlings are shipped, the emphasis may shift to rationing out the residual forage to supply the nutritional needs of a cowherd in the most economical manner. At the same time, you can be using the cows as tools to condition the forage so that its quality and health will be high when the next turn of stockers arrive. The plan must account for the needs of all of the animals that are present or will be present during the grazing season in an economically feasible manner that promotes the health of the entire soil-plant-animal complex.

The cardinal sin in grazing management is to become mechanical and inflexible.

As circumstances change, grazing management practices will change to make the best of the situation. If the desired conditions can be visualized, plans can be formulated to create these conditions; a logical plan can be likened to a road map that points the way to a desired result. Humans are not capable of keeping track of and making logical use of all of the information that influences a complex operation unless that information can be applied through a logically designed plan. Just as no person can keep all of the needed facts in his or her mind, no plan can account for all of the possible changes in all of the factors that ranchers encounter daily. The solution is to gather all available, useful information and use that information to make the best possible plan to achieve the desired results. Understand, however, that the plan is not perfect and will require changes as conditions change; no plan will ever be perfect, but monitoring for results will allow a plan to be improved—and even a poor plan is much better than no plan.

The plan may be useless, but the planning is invaluable.
— General Dwight Eisenhower

To me, planned grazing means planning to locate grazing animals in the time and space so as to best meet the needs and goals of all parts of the grazing operation: ecological, financial, and human. This goes far beyond the practice of rotational grazing. Done right, planned grazing will be designed and implemented to promote the well-being of the animals, of the land or soil-plant-animal complex, and of the graziers. The plan will account for the constant changes in everything, from growth rate and nutrient content of forage in various areas, to nutritional needs of assorted classes of animals, to market situations and financial conditions. All of these factors are constantly changing, and if the plan is to succeed, all must be monitored constantly and the plan updated to account for the new conditions. This sounds like a tremendous amount of work and extremely complicated, but in practice, it can greatly reduce the amount of time, labor, and money spent and can make good management much easier to achieve. The rationale for this lies in the fact that all parts of the operation (the animals both wild and domestic, the grasses and the forbs, the soil organisms, and the people) are all part of a functioning "whole," and anything that does long-term harm to one part damages the whole. We would do well to start our management planning with the old advice to physicians: "First: do no harm." This sounds simplistic, but it is not; the way to build and maintain a grazing operation that is profitable and stable for the long term is to manage so as to build the health of all parts of the operation.

Good advice for physicians and managers
"First: do no harm."

Manage for What?

There is no doubt about the power of grazing as a force to change the environment; poor grazing management has degraded vast areas of the world's grasslands, and improper grazing is second only to the plow as a force for destroying grassland. Grazing is an extremely powerful tool, and as with all tools, its effects can be destructive or beneficial depending on how it is used. Blanket condemnation of grazing because it has been and still is misused is as wrongheaded as denouncing fire because it can burn down houses. Planned grazing management utilizes human intelligence in the manipulation of grazing animals and their effects on the local environment for the purpose of effectively converting solar energy first into biological energy and then into valued products. These valued products consist of the obvious meat, milk, wool, and so on, but they also include other benefits that may not be so readily apparent.

Correctly managed grazing is an effective tool that can be used for the following:
- Restore damaged grasslands
- Improve soil stability and productivity
- Increase profitability of livestock operations
- Improve the hydrological characteristics of watersheds
- Increase biological diversity
- Improve wildlife habitat
- Increase production of crops, livestock, and wildlife
- Promote financial stability within an area
- Improve quality of life for the people living on the land

The basis of planned grazing management is the synchronization of the amount, timing, and quality of the available forage with the needs of the animals being grazed while improving the health of the whole soil-plant-animal complex.

Factors Controlled in Grazing Management

While grazing management can be used in combination with many other tools, only a few factors are actually involved in the management of grazing. Although grazing management deals with many and complex factors, its benefits are achieved by manipulating relatively few factors. Everything done in planned grazing is done by managing the following eight factors.

1. **Stocking rate**: *The number or pounds of animals stocked per unit of grazing land for the grazing season.*

    The stocking rate of an area of land is a direct reflection of the amount of forage available in that area during the time the animals are present. Overstocking is common and is second only to continuous grazing as a cause of ecological and financial damage on grazing operations.

    A. Stocking rates that are too high cause overutilization of the forage, which, in turn, causes the following:
    a. Poor animal performance on all fronts; slow growth rates, low reproduction efficiency; and increased disease and parasitism due to poor nutrition and contamination with waste products
    b. Slower forage growth because of the reduction in amount of solar energy converted to biological energy due to reduced leaf area
    c. Reduced ground cover and plant canopy, with resulting poorer water cycle and a loss of soil life due to drying and to extreme temperatures

d. Shift in forage species toward plants that can protect themselves from overgrazing, resulting in loss of good forage species and an increase in poisonous plants and low forage–value plants
e. Reduction in the amount of forage grown and consumed, and loss of soil life, which decreases the amount of minerals in cycle
f. Increase in need for supplemental and substitution feeds, and higher costs
g. An unstable situation both ecologically and financially, with little ability to withstand either environmental or financial stress
h. The effects of articles b through e are cumulative, and the result of prolonged overstocking can degrade productive grassland in a relatively short time

B. Stocking rates that are too low cause the following:
a. A reduced percentage of plants being grazed or browsed, which means the amount of old and ineffective plant material increases, and efficiency of solar energy capture and conversion drops
b. Very low stocking rates will cause a shift in vegetation types and ground cover; in arid areas, the shift will be toward bare ground and xeric plants, while in humid areas, the shift will be toward forest
c. Low stocking rates bring reduced animal numbers and reduced profitability, as overhead costs are concentrated on fewer animals

d. Good individual animal performance, but low production per unit of land
e. Low stocking rates will not stop overgrazing; the number of plants being overgrazed will be reduced, but the number of plants suffering from lack of grazing will increase. Prolonged lack of grazing is damaging to forage plants, with the most productive species most severely affected

Finding and maintaining an optimum stocking rate is one of the most important jobs of a good manager. All factors of supply and demand must be considered when setting a rate, but especially important is the reliability and predictability of forage growth. Set the stocking rate, the demand for forage, to be in line with the amount of forage that can reasonably be expected to grow in less than ideal growing conditions, and be prepared to reduce forage demand as soon as poor growing conditions develop. Reserves for use during seasonal periods of slow growth and during drought periods should be provided by leaving more residual on the grazed plants rather than by setting aside areas of forage. Forage that is set aside will soon slow in growth rate and will lose quality over time, while additional leaf left on the grazed sward will increase rate of growth and add to the overall supply. It is feasible to stockpile forage toward the end of the growing season for use during the dormant season if care is taken that the forage is of sufficient quality to be useful. Warm-season forage grown in June will have very little nutritional value left in it by January; warm season forage has the greatest value during the dormant season when it has been lax grazed under high stock density during the growing season.

Thousands of acres of native bluestem grass are burned off every year when by, rationing out the forage using high stock density, millions of cow days of forage could be gained; grazing this forage at high stock density would provide most of the benefits of weed control and quality early growth sought with burning and none of the downsides of burning. Cool-season forages can be stockpiled to good advantage starting in the early fall; tall fescue is particularly valuable for this practice.

If it becomes necessary to feed energy (hay, grain, silage, etc.) during the growing season, the stocking rate is too high; in most cases, feeding energy at any time except when grazing is impossible due to ice, deep snow, flooding, and the like, is a signal that the stocking rate is too high or that grazing management is poor. Substitution feeding of stored or purchased feeds instead of pasture is a profit killer.

When a cow is out grazing, she is working for you; when you are feeding a cow, you are working for her (I know I said this before, but it is important).

If your operation routinely uses large amounts of hay or other outside feeds, it would be wise to consider methods of adjusting to your environment; reduction in stocking rates should be one of the first options considered. Over stocking is the most common cause of poor economic performance of grazing operations.

The financial growth engine of ranching is the ability to convert something of low value (forage on the land) into something of high value (beef, wool, wildlife, etc.).

Substituting costly feeds for low-cost feed does away with this financial advantage. Supplemental feeds such as minerals or protein can be beneficial and profitable when needed, but the vast majority of the total ration should come from grazed forage.

Your resources are grass, livestock, and money; it is hard to have too much grass or money but real easy to have too much livestock.
-Bud Williams

2. **Stock density** *is defined as the number or pounds of animals present on a given area at a given time.* Stock density can be increased by either increasing the number of animals present on an area or reducing the size of the area being grazed.

Stock density can be managed very well by herding; in France, sheepherders whose flocks graze in the mountains move them 8–12 times per day to different types of forage in a sequence that experience has taught them will maximize forage intake and animal performance while making the best use of the available forage. Most of us will use fencing to subdivide areas and control the location in time and space of our animals; subdivision will be discussed in the land planning section, but it is obvious that increasing stock density requires adding to the number of subdivisions utilized by each herd of animals.

The difference between stocking rate and stock density is a function of time and concentration. High stock density grazing consists of placing a large number of animals on a small area for a short period of time, followed by an appropriate recovery period in which no grazing animals are present. In the following example in figure 2, placing all one hundred animals in one of the ten equally sized subdivisions increases stock density tenfold in the area where the animals are concentrated. It also decreases the stock density to zero in the areas not being grazed and provides the plants in these areas time to recover from being bitten, also giving soil organisms time to process the feast of nutrients in dung and trampled vegetation. In the example, assuming that the subdivisions are equal in size and that animals stay in each subdivision for the same length of time, 90 percent of the total area will be resting from grazing at any point in time. When animals are concentrated to this extent, the percentage of plants impacted by the animals (bitten, trampled, or covered by dung) increases greatly. Spreading the animal impact in this manner brings more plants into the equation so that all or most are utilized without any being overutilized. Being bitten is a positive thing for a forage plant, provided that the action is not repeated too quickly; likewise, the deposition of dung and urine, the trampling, and the disturbance of the soil surface by hooves is beneficial for the soil and vegetation if it is not too soon repeated. There is a vast difference in effects on the soil, the vegetation, and the grazing animals between twenty animals grazing one hundred acres for fifty days and the same twenty animals grazing the same one hundred acres fifty days at two acres per day. When animals are continuously present in an area, they will always degrade favored areas; the effects of normal activities such as grazing, trampling, and dumping of body wastes that are applied repeatedly to the same area without respite are destructive to soil and to vegetation. The effects of the same activities become beneficial if they are concentrated in an area for a short period of time and the area is then rested from these activities.

Unlike stocking rate, stock density becomes more valuable to the forage and the soil as its rate of intensity is increased. A number of ranchers in various parts of the world stockpile forage and use multiple stock moves per day to apply stock density rates of 250,000 pounds or more per acre. The results on the soil and the forage of this technique, commonly called mob grazing, have sometimes been phenomenal, with large increases in soil organic matter content, soil life, and forage growth. The practice seems to yield the best results when forage is allowed to grow and accumulate to full maturity; animal performance is maintained on this rank forage by putting a tremendous amount of forage on offer, allowing the animals to select only the most nutritious portions and trample the rest into the soil. The ratio of eaten to trampled forage that most good operators strive for is about 40 to 60 or even 30 to 70; animal performance will fall off if animals are forced to consume a larger percentage of the older forage. The practice is labor intensive if multiple moves are made per day, and great care must be taken to not stress the animals by failing to allot sufficient forage, but the results in improving soil productivity can make this an extremely valuable technique, given the right conditions. *It is not something that beginning graziers should attempt until they gain experience.* Ian Mitchell-Innes of South Africa has been a leader in developing and refining this practice.

The stocking rate in figure 1 is ten cows on one hundred acres, or one cow per ten acres, while the **stock density** is also one cow to ten acres, because all ten cows have access to the whole one hundred acres.

```
        100 acres 10 cows
              C C      C C
      C C           C C
            C C
```

Figure 1

The stocking rate in figure 2 is also ten cows on one hundred acres, or one cow to ten acres.

When all ten cows are placed on ten acres, the **stock density** becomes one cow per acre on that ten-acre tract. On the other tracts at this point, the stock density is zero.

Figure 2

A. Low stock density with long grazing periods results in the following:
   a. Increase in overgrazed plants, because with long grazing periods, animals concentrate grazing pressure on regrowth of previously grazed plants, causing a decrease in plant production due to reduction in both energy flow and amount and rate of mineral cycling
   b. Increase in amount of senescent plant material since with a large area of forage on offer, many plants are not defoliated early in their growth cycle and thus lose quality and are refused by the animals. This results in decreased plant production due to reduction in both energy flow and in the amount and rate of mineral cycling, because the older plant material is less efficient
   c. Good individual animal performance, because animals are able to select from a lot of forage

but low production per unit of land, because fewer animals are present

d. Some spots are overutilized and degraded while other areas are utilized very little; both conditions favor a shift away from healthy grassland, with increased encroachment of weedy plants

B. High stock density results in the following:
  a. Uniform utilization of the vegetation, with all or most plants being bitten. This will result in a forage sward that is fairly uniform in physiological age; the age (state of growth) of the forage can now be manipulated so that the characteristics of the forage presented to be grazed matches the needs of the animals. Different classes of animals have different requirements for protein and energy; the closer the match between these needs and the content of what is presented, the more value is derived from the forage
  b. As high stock density requires short grazing periods, the overgrazing that results from the grazing of regrowth can be halted by moving before plants develop significant regrowth. Providing recovery periods of the proper length can halt the second type of over grazing that results from grazing plants before they are fully recovered
  c. By controlling the degree of forage utilization and the timing and length of the recovery periods used, it is possible to favor the productive forage plants and shift the level of biological

succession toward a level of more stability, greater diversity, and higher production

d. Grazing uniform-aged forage in a short period of time increases animal performance by providing a ration that is correct in the balance of protein and energy for the class of animals being grazed; short grazing periods keep the quality of the forage the same from day to day, and this lack of variation improves rumen function and animal performance

3. **Stocking mix**: *The number and pounds of each type of animal in the total of all animals being grazed.*

An important factor in both stability and profitability of a grazing operation will always be how well the available forage suits the needs of the animals being grazed. Diet preferences differ by species: cattle, bison, and elk prefer grass; sheep and deer prefer forbs, and goats prefer browse. These preferences have foundations in the basic physiology of the animals; although all animals will consume a mixture of the available plants, they do best when the majority of their diet consists of their preferred foods. Preference entails more than palatability; preference goes beyond taste to include the total well-being of the animal. Animals prefer certain types of food because they have the physical and physiological attributes to harvest and utilize these foods and because these foods meet their nutritional and metabolic requirements. Animals know when the available diet does not meet their needs and will do their best to change the situation. Cattle fed a high grain ration will resort to eating wood in an attempt to gain the fiber they need, and goats on lush grass pasture will become escape artists in the attempt to find the growing points of woody plants that they need. When any animal is forced to live on an unnatural diet, performance will suffer.

Poor performance can be caused by animals being forced out of their normal grazing technique. Sheep, goats, and deer are selective grazers that select a very high-quality diet one leaf at a time and are at a disadvantage in forage swards that lack diversity of plants and thus limit their ability to select. Goats have little genetic resistance to internal parasites, because they evolved in dry areas browsing on wooßdy plants well up off the ground where parasites were rare; when they are forced to graze at ground level, they quickly become heavily parasitized by the large number of infective larvae that exist on the first few inches of vegetation. Cattle, bison, and horses are mass grazers that feed by coming in over the top of forage and tearing off the top portion, and they do best when forage is dense and abundant. All animals forced to graze short forage have more trouble with internal parasites than those grazing taller forage.

The decision of which species and in what proportions to stock depends in part on the vegetation available in an area. In areas with a mixture of grasses, forbs, and browse, it is often possible to produce dramatically more total animal product by stocking with a mixture of cattle, sheep, and goats than would be possible by stocking only one species. When all parts of the vegetation are being utilized, the amount of solar energy captured and converted will be higher, and the amount and rate of mineral cycling will be higher; these improvements will lead to increased soil and plant health, which will contribute to both stability and profitability. Whether stocking with multiple species is a valid practice will depend not only on the feed available but also on other considerations such as fencing costs, knowledge, availability of markets, and personal preferences. There is a learning curve in going from cattle only to adding sheep or goats, but in the right circumstances, sheep or goats can be a valuable addition to an operation.

A second decision in stocking mix is which classes (grown cows, stockers, weanlings, and so on) of animals should be carried and in what proportions; the primary determining factor for this decision is the availability of quality forage both within and between years. Different classes of animals have differing nutritional needs; young growing animals have a higher requirement for protein than do mature animals, while lactating or fattening animals have a higher requirement for energy. The stocking mix should reflect the characteristics of the available forage over the grazing season. Classes such as dairy cows or animals being fattened for grass-fed beef require long periods of green forage and would be a poor choice for areas where the green season is short due to moisture or temperature. A better choice for an area of short green season might be stocker animals that would be present only while quality forage was available or cows that would calve on green grass and maintain themselves the rest of the year on lower-quality forage. In areas prone to drought, a combination of these two classes can give flexibility, with stockers moved early or not purchased in dry years.

4. **Recovery period:** *The amount of time between grazing periods in which a given area is free of grazing animals, and the timing of these periods.* The amount of time required for complete forage recovery is determined by the rate of growth of the forage.

    Defoliation of forage plants by grazing animals is natural and beneficial to both plants and animals if sufficient relief from grazing pressure is provided for the plants to recover from its effects. Conversely, continuous exposure of forage to grazing is damaging to the forage as well as to the soil and animals. Damage done to forage plants during continuous grazing occurs in two ways: (1) overgrazing when a plant is re-bitten before it has had sufficient quality growing time to recover from previous defoliation and (2) underutilization when a plant is not defoliated for prolonged periods and shades out its own growing points. Continuously grazed areas, which are stocked at reasonable rates, will always have a lot of senescent forage plants because all plants in the area are on offer at the same time and many will not be bitten early in the season when forage is most plentiful. The damage from overgrazing is prevented by removing grazing animals from the area until the grazed plants have fully recovered from the effects of being grazed. This point occurs when the material consumed by animals or broken off has regrown and has been photosynthetically functional long enough to manufacture at least the amount of energy required to regrow the lost material.

The amount of time required for recovery from grazing is determined by: the quality of the growing conditions, the amount of material removed, and the physiology of the plant in question. In a perfect world good grazing management (management favorable to the soil, to the forage, and to the animals) can be achieved by grazing a portion (the amount will vary according to the timing of and purpose of grazing) of the leaf area of each plant in a short period of time and then removing the animals until the grazed plants have fully recovered and are ready to be grazed again. If the timing is correct, the forage can be grazed when its total nutritional value is highest for the class of animals it is feeding. The length of recovery periods can be varied to meet the needs of special situations; recovery periods might be shortened to provide young growing animals with a ration higher in protein, lengthened to provide fattening animals with more energy, or lengthened still more to stockpile forage for the dormant season. Just be aware that all actions have consequences, and monitor the results of your practices. Forage stockpiled by long recovery periods can be used to feed animals during the dormant season and can also be utilized with very high stock density to increase the amount of carbon deposited into the soil by intentionally trampling a lot of forage into the soil. Close monitoring of animal well being is critical when grazing fully mature forage; only a portion of this forage will be high quality and animal performance will suffer if animals are forced to consume material that is old enough to have lost digestibility.

The recovery periods should always be long enough to allow the forage sufficient time for complete recovery; the exception to this rule is when forage is intentionally overgrazed to weaken it, to allow interseeding of other species. This is a powerful tool, and its use should be carefully considered and its effects closely monitored. Proper grazing builds healthy grasslands and profitable ranches; conversely, improper grazing degrades grasslands and reduces profitability.

The amount of time required for a plant to fully recover from being grazed depends on the quality of the growing conditions, the percentage removed from the leaf area of the plant, and the genetic growth characteristics of the grazed plant.

A. The quality of growing conditions at a given point in time is determined by the following:
   a. Amounts and balance of soil water and soil gases
   b. Soil productivity: depth, water-holding ability and fertility—amounts, availability, and balance of minerals
   c. Amount and health of biological activity in soil

    d. Amount of sunlight
    e. Soil and air temperatures
    f. Area and physiological state of leaf per plant
    g. Amount of stored energy per plant at the start of growth
    h. Presence or absence of insects and disease pests

The more material removed by grazing, the longer it will take a plant to regrow; leaf must regrow before it can replace the lost energy. For many of the high-quality forage plants, when more than half of the leaf area is removed, the roots cease to grow. If the defoliation is severe, the energy required to grow new leaf area will be transferred from the root area, causing the death of roots. If roots are severely impacted, the needed recovery time will be significantly increased, because both leaves and roots must be replaced for full recovery to be achieved.

The physiological makeup of the grazed plants is an important factor in the amount of time required for the plant to recover; in general, the tall-growing, highly productive, perennial plants require more time to recover than do shorter, less-productive plants. The amount of recovery time required is closely related to how fast a forage plant reaches maturity; slow-maturing plants such as Big Bluestem and Indiangrass require more time to recover than do quick-maturing plants such as tall fescue or Bermuda grass. Slow maturity is a hallmark of forage quality, as slow-maturing plants stay in phase 2 (the stage between full leaf functionality and beginning of leaf senescence), in which plants are most nutritionally valuable, for a larger percentage of their life cycle. Using short (or no) recovery periods discriminates against the valuable slow-maturing plants and, if continued, will cause a shift to the early maturing, less productive plants.

Setting optimum recovery periods is a prime skill of the good grazing manager. Initial ballpark recovery periods can be set based on experience, but these must be under constant review and adjustment. The only valid way to set recovery periods, except when stockpiling forage, is according to how fast grass is growing; as the factors that control growth rate are constantly changing, the manager must monitor growth rates and adjust recovery periods to suit conditions. If there is question regarding how fast grass is growing, flag a grazed plant, record the date and the height of the plant, and come back to that plant periodically to check growth rate.

Example: If the plan calls for forage to be twelve inches high at turn in, and a thirty-five-day recovery period is being used, the forage will need to average .2 of an inch of growth per day to be on schedule after forage is grazed to a height of five inches (12"- 5" = 7"; 7"/35 days = .2 inches/ day). If the forage is not growing as fast as needed, then the plan must be adjusted to yield the desired results. Numerous reasons exist for the difference between what was planned and what is happening, but they all come down to how fast the forage is growing during this period of time; it may be that thirty-five days is not enough recovery time for the class of forage being grazed, or it may be that five inches of residual does not leave enough leaf area to promote the rate of growth needed or that the quality of the growing conditions is not as good as supposed. If the problem is too little residual being left after grazing, it is likely that the stocking rate is too high, and either the area being grazed needs to be increased or the number of animals needs to be reduced. If the amount of time required for the forage to recover has been underestimated or the quality of growing conditions has been overestimated, the recovery periods must be lengthened. The point of this example is to stress that recovery times cannot be set on a calendar basis and be expected to succeed; all decisions in the grazing plan must be created based on expected conditions and modified according to the actual conditions.

The grazing plan is based on your best guesses as to what will happen; if you have ever made a mistake, you might better monitor and adjust.

---

5. Grazing period: *The length of time that a given area is grazed between rests, and the timing of these periods.*

The length of time that animals are held on an area of forage strongly affects animal performance. Animals are exactly like people—they always eat the best first, and they expend a lot of time and energy looking for another bite as good as the last. Every hour that animals are present on an area, both the quality and quantity of the forage on offer decreases. Leaf and other high-quality material is consumed first, so the digestibility and feed value of the forage on offer decreases as forage becomes scarcer due to consumption and as spoilage from trampling and dunging increases. In continuously grazed areas, the amount of forage spoiled by dunging and trampling can easily be 30 percent or more.

As the top portions of the plants, which contain a lot of sugars and other easily digested starches are consumed, the remaining material contains increasing amounts of cellulose and less digestible carbohydrates and lignin. The types of rumen flora that are able to utilize the abundant sugars and starches are different from those that can utilize high cellulose, which are different from those that can break down more woody material. As the content of the diet changes, the proportions of starch digesters to cellulose digesters to woody digesters in the rumen bacteria population changes to reflect the content of the ingested material. The longer the graze period (up to the point at which regrowth becomes a significant part of the available forage), the bigger the shift in diet content toward woody material, and the more radical the shift in the rumen bacteria toward the species capable of using the lower-quality material. When the animals are moved to a new break of forage, the diet suddenly changes back to a high sugar and starch content, and the rumen bacteria population is again wrong for the diet being consumed. Anyone who has fed livestock in a confined situation knows that rule 1 is to not make sudden changes in the animals' diet. A sudden shift in the composition of the diet brings about digestive upsets and poor animal performance; this is especially true of ruminants that rely on rumen bacteria to break down the material they consume.

From the standpoint of ruminant animal performance, maintaining day-to-day uniformity in the quality of the diet is critical

The rumen flora can adjust to make good use of some rather low-quality material, provided that the quality of the diet is not constantly changing. One reason some grazing research projects have shown poor animal performance is that the projects were designed and operated with long graze periods. Short graze periods are one of the simplest ways to increase animal performance. Regardless of what percentage of the forage in an area is to be consumed, doing so with short graze periods will yield better animal performance. A five-day diet of one-fifth ice cream and one-fifth cardboard box that it came in yields dramatically different results from a four-day diet of ice cream followed by a one-day diet of cardboard. Most of the improvement will come from the uniformity of the diet allowing the rumen flora to stabilize and do a better job of digesting the forage, but improvement will also come from increased forage intake brought on by a change in animal behavior. A fresh break of forage is a treat to grazing animals, and even animals that are full and have stopped grazing will start grazing again when offered fresh forage in a competitive situation.

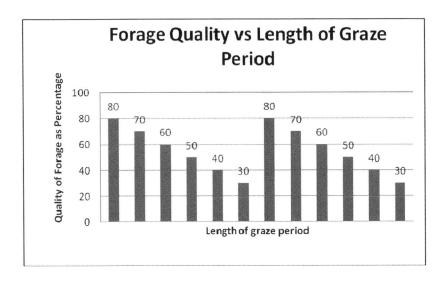

6. Animal impact: *The effects on the soil-plant-animal complex due to the presence of animals.*

Animals create change in their environment simply by being present and carrying out their life processes. For grazing and browsing animals, the most visible changes relate to the material consumed by the animals. By consuming, digesting, and excreting plant material, animals change the physical, biological, and chemical composition of their environment. These changes can be dramatic, as when elephants break down trees and increase the ratio of grass to timber, but most changes are more subtle. When animals consume forage, they change the status quo. Defoliated forage plants are stimulated to grow new foliage; at the same time, as a part of their mechanism of self-protection, they are prompted to release root exudates into the soil, which stimulates soil microbes and improves growing conditions for the grazed plant. Removing top growth opens the plant canopy to sunlight, which can stimulate the germination of dormant seeds and increase the seedlings' chances of surviving to become a viable plant. By putting plant material through their digestive tracts, animals hasten the biological breakdown of the material, increase the rate of mineral cycling, and prevent the loss of nutrients to chemical breakdown.

Animal impact can also have effects that are not beneficial. Continuous grazing is always deleterious to some extent. Preferred plants can be overgrazed to the point that they are eliminated and replaced with lower-quality plants that can withstand abusive grazing or with plants that are not normally grazed. If grazing is too severe or repeated too often, it becomes a negative influence on all of the biological processes, and the health of the land suffers. The effects of grazing animals can be either beneficial or damaging, depending on how the grazing process is carried out. I think it is enlightening as to how well we understand the process that the plants that we term improved forage plants are improved in the sense that they can withstand abusive grazing and can use lots of nitrogen fertilizer.

Animals impact their environment in numerous ways aside from consuming vegetation. Although all animals, even mice, form trails when they move back and forth repeatedly over the same route, hoofed animals are exceptionally prone to forming these vegetation-free paths. Trails are formed by the compaction of the soil that occurs when the weight of the animal is concentrated on the edges of hooves and this action is repeated on a regular basis. This compaction prevents both water and air from entering the soil and creates a hostile environment for plant growth; this hostile environment, combined with the damage to plants caused by repeated trampling, can prevent all or most plants from growing in the trail. The bare soil of the trail then becomes a starting point for soil erosion. Similar damage will occur when an area such as a watering point, shade tree, or feed ground is in continuous use. A compacted layer of soil or hoof pan is common in all areas that are in continuous use by animals; the natural processes that relieve soil compaction, such as actions of soil-dwelling organisms, root growth and expansion, and contraction and expansion from soil moisture changes, can be overwhelmed by repeated pressure from walking animals. Although it seems illogical at first glance, concentration of animals on a small area for short periods of time causes less soil compaction than does having the same number of animals continuously roaming the entire area. Under high stock density, compaction will take place in the grazed area, but because periods of high concentration are followed by periods with no animals present, the natural processes have time to relieve compacted soils. The exception to this is very sandy soils that tend to fluff up under continuous animal presence and become firmer under high stock density.

7. Herd effect: *The effects on the land and vegetation brought about by animals in an excited herd mode.*

The behavior of a herd of excited animals is quite different from that of animals that are not excited. When a herd is grazing quietly, each animal carefully places its feet, avoiding dead grass, manure pats, bushes, and anything else that might cause it to stumble or expend unnecessary energy.

If the same herd is frightened or is excited by competition for a treat (such as cake or alfalfa hay thrown around), foot placement is forgotten, and the results to the land are completely different. It is possible to use this trait in positive ways; by tossing flakes of high-quality hay around the head of the gully, beginning gullies can be healed by inciting cattle to break down steep gully walls and trample vegetation into a soil holding mat. The same tool can be used to break up dense thickets or to trample patches of moribund forage into the ground so that biological decomposition can begin.

Herd effect can also be destructive; I was called to a ranch several years ago, because they were having trouble controlling their cattle with electric fences. What I found was a classic case of the difference between how cattle react when in a normal mode and how they act when in an excited herd mode. It was winter, and the cattle were being fed hay and cubes for their total ration; the cattle were in perhaps a dozen herds strung out along several miles of highway and were being fed on feed grounds just off the blacktop. All of the cattle would be on the feed grounds when the crews started feeding, and as the hay trucks passed on the road, the animals would attempt to follow, get excited, and run through the electric fences separating the herds. As the animals would never break through an electric fence when not excited, the solution was simple. I had the feeding crew start by feeding the herds closest to the hay barns first and do it quietly so that they were not tormenting the hungry cattle by parading hay by under their noses. By the third morning, the cattle were contentedly waiting for their turn to be fed. It is far too common that animals being fed supplements perform poorly, because the manner in which the supplements are fed creates stress on the animals.

8. **Animal behavior:** *Animal actions and reactions that can be understood and used to bring about desired results.*

As mentioned earlier, the diet preferences of animals differ according to species, but diet selection is also affected by preferences of individual animals. You probably like what your mother fed you as a child, but you also tend to like what your mother ate when she was carrying you in her womb. Animals are no different; they prefer familiar foods. When a ewe eats sagebrush during her pregnancy and lactation, her offspring will have a lifelong taste for sagebrush. The opposite is also true; if a cow eats only grass, her offspring will be unlikely to learn that some forbs and browse are good to eat unless they are forced to eat these things because of a shortage of grass. Even then, they will be unlikely to continue to eat them when grass is again abundant, as they learned to consume them only out of necessity during an unpleasant time. If, however, animals are introduced to exotic foods in a positive manner and do not suffer a negative reaction, they can and will add to the number of plants that they routinely utilize. In areas with mineral deficiencies in the soil, animals born and reared in the area learn by positive feedback which plants have the highest content of the scarce mineral and seek out these plants.

As Dr. Provenza and his crew at www.behave.net have shown, the antiquality factor in one plant can be the antidote for an antiquality factor in another plant; when animals learn the proper combinations, they have the advantage of increasing their inventory of edible forage. The kind and amounts of antiquality factors vary by plant species as well as by season and stage of growth; a plant that is not utilized when immature may be relished when it matures, and vice versa. When animals are moved to an environment different from the one in which they were reared, there is almost always a drop in animal performance until they learn which plants to eat and in which combinations. The more unfamiliar the environment, the longer this process will normally take. (Kathy Voth has and is doing some excellent work in teaching animals to consume and benefit from weedy plants that they normally would not eat. Check out her work at www.livestockforlandscapes.com.

All of the domestic grazing animals are herd animals. The herd is the basic social unit of these animals, and all members of the herd quickly learn their position in the social hierarchy; being secure in this position and knowing how to react to both social superiors and inferiors gives them a sense of safety and gratification. Anything we do to cause animals to lose their sense of decorum, such as pushing a lower-ranked animal into the space of a higher-ranked animal, is disruptive and stressful.

The most stressful situation for a herd animal is to be isolated; whenever we pen a single animal, we are inviting all types of stress-related maladies, from disease to dark cutters. Years ago, I was weaning a set of calves on a hotwire in the pasture when a single calf got out of the calf group but did not get in with the cows. I saw the calf early in the morning frantically trying to join one group or the other, but I did not go out and help it until late afternoon; by the time I penned the calf, it had a full-blown case of respiratory disease. This was the only case of respiratory disease I treated in the several hundred calves that I weaned that year. We cause a lot of animal performance problems by not recognizing normal animal behavior. Animals will always tell you what they want to do; the secret to stress-free handling is to plan your actions so that what the animals want to do is what you want them to do.

It should be a lot easier to outthink animals than to outmuscle them; if you find that the opposite is true, perhaps you should consider a career in professional wrestling or Chicago politics.

An example: If animals go through a gate located in such a way that the first animals through turn and come back toward the herd on the opposite side of the fence, the animals following the leaders will turn back before they get to the gate. When planning fencing, a prime consideration should be how animals will react in a given situation. A little thought and time spent in designing pens and fencing can result in reduced stress and ease of handling. It is a lot easier and more cost effective to fix the mudhole in the gate than to stress a bunch of cattle and lose your religion trying to force cattle through the mud.

Purpose

The purpose of planned grazing management is to bring about a specific set of desired results.

A ranching operation or any other land use endeavor involves interactions between (at the very least) people, land, animals, plants, and money. As every action we take affects all parts of our local environment as well as the people and animals dependent on that environment, goals (desired results) need to be established for all parts of the operation; we need to spell out the conditions that we desire to see in each part of our operation so that we can judge the effectiveness of the management techniques we employ. Many of the problems of modern life (political, social, and agricultural) result from focusing on one aspect of an endeavor rather than on the health of the whole.

For a ranching operation, the elements that must be considered when setting goals are the following:

- human needs and desires
- needs of the land
- needs of the animals
- needs of the operation as a whole
- financial needs

- The human needs and desires of the people involved must play a large role in establishing the things that the operation is to accomplish.

What do you want? You get up early, work until late, agonize about the well-being of your stock as though they were your kids, and fight the constant money problems. Why put up with all of this when, with the same dedication and work ethic, you could go to town and make a lot more money with a lot less worry? You do it because you want something more than money. You want to build a home for your family and provide them with a means to live through your and their own efforts rather than according to the capricious decisions of some bureaucrat or corporate VP; you want to see the land that is in your care grow healthier instead of sicker. You want to see God's handiwork up close and share this experience with your loved ones; you want quality of life. I will say again, ranching is a business and must be operated as a business; just don't forget the reason for having the business!

We are human, and our strongest motivation is our own well-being and that of our families; at times, we are tempted to do things that offer short-term benefits but must be paid for with long-term costs. (You know, like when politicians spend money that will have to be taxed from our grandchildren to buy voters with promises of free goodies.) To me, hell on earth would be to have to tell my grandkids, "This was a good place when I got it, but it is kind of run-down now, because I needed money more than I needed a healthy ranch." The old saying, "You can't starve profit out of a cow" is true; abusing your resource base for short-term financial gain is a sure way to destroy quality of life for you and your heirs.

Quality of life means different things to different people, but most share some of the same desires; we want physical and financial security for ourselves and our families, and we want the time and the means to practice our religion and to enjoy hobbies or activities with friends and family. One of my goal that I rarely mention is to never do anything that is contrary to my principles; when I look in the mirror to shave, I want to be able to look myself in the eye.

- Needs of the soil-plant-animal complex we call the land. What must be done to preserve and build biological capital.

The complex that we call the land is an intricate association of uncountable and widely diverse organisms utilizing and replenishing the inanimate resources of an area in a never-ending cycle of birth – life – death – decay – birth. For the vast majority of the time that mankind has been on the earth, humans had no more lasting effects on their environment than did a similar weight of other omnivores such as bears. At some point, though, human intelligence developed, and man became the most dangerous animal on earth. With intelligence came tools and fire and agriculture and domesticated animals, and man's influence on the environment was magnified ten thousand fold. Even then, the number of humans was so small that mankind's effects were insignificant in relation to the size of the earth. Today, the number of people has exploded, and our tools are far more powerful; it is completely possible that we could overwhelm the ability of the natural world to recover from our abuse. I am not talking about carbon dioxide from fossil fuels causing global warming; that is a scam promoted for political and financial advantage and is not based on science. I am talking about the worldwide destruction of life and the means to support life; this is being done out of ignorance, but it is also being done by design. I am talking about the desertification of millions of acres of grasslands through abusive grazing and inappropriate farming and about the destruction of ocean life through excessive fishing and pollution. I am talking about farming practices that destroy soil organic content and the soil life that is the basis of soil productivity, and I am talking about the ever-increasing amounts of increasingly powerful poisons that are being routinely used in all aspects of our life. I am talking about the industrialization of agriculture for political reasons and the resulting market disruption and destruction of local rural communities. We have become extremely dependent on a very small number of food crops grown in degraded soil, and these crops come from very scant genetic bases; one newly mutated disease of rice, corn, wheat, or soybeans could cripple commercial food production worldwide and trigger starvation on a scale never before seen.

Why am I spending so much time on a subject seemingly so far removed from grazing management? All of the problems detailed above have their roots in the loss of biological diversity and biological capital. The most powerful tool available to us to increase soil health and biodiversity and to reduce the need for poisons and acid salt fertilizers is planned high stock density grazing.

The greatest mistake made to date by modern agriculture, and there have been some doozies, is to take animals out of crop production.

Without animals to break down crop residues, build soil health, and give monetary value to soil-improving crops, modern agriculture has become ever more dependent on purchased inputs and more at risk to drought, disease, and weeds. An agricultural system that relies on animals to build soil fertility for crops can be at least as productive as what we have today, and far more profitable and stable.

Building all aspects of biological capital must become a primary goal not only for ranchers but for society at large.

- Needs of the livestock. How to provide all that is necessary to maintain healthy and productive animals

We must of course provide for the nutritional needs of our animals. Ample clean forage of the kinds and maturity proper for their physiological state that can be harvested with a reasonable amount of grazing work is basic; as is, ample clean water, and access to any needed supplements such as salt and minerals. In addition, we know that at times animals will require shelter from the elements, and a part of the grazing plan should be where animals are to go when the weather gets rough, what they are going to eat while they are there, and how we are going to get them to shelter. If you have ever tried to drive a set of cows into the teeth of a fast developing "blue norther" because the only wind break is north of you, you understand.

If animals are upset, something is wrong and since you are in charge, it is your fault

---

I was called to a ranch once, because it was losing baby calves at an unacceptable rate. The owner was not there, but when I met the foreman and asked about the problems, the cause of the problems became clear. The ranch had recently changed to a rotational grazing program, and the foreman didn't like it, didn't believe in it, and was sure it wouldn't work—and he was right. I don't know whether it was intentional, but the foreman was doing everything he could to make sure this foolishness failed; he gave me a litany of things wrong with the program: the herds were too big for the gates and calves were being trampled to death in the mud during moves, it took too much time to be moving cattle all the time, the cattle were losing weight, and so on. It didn't do any good to point out that he was keeping the cattle too long in each paddock and they were running out of grass and were frantic to get to a new paddock. Moving the cattle before they ran out of grass would solve the poor animal performance and trampled calves, but it also wouldn't hurt to make the ten-foot gates wider and fix the mudhole that was in every gate on the ranch. Regardless of what ownership and management think, the people on the ground have to believe in the program, or it will not work.

When they are confined, the well being of animals must be monitored constantly. When I was at Texas A&M many years ago, someone shut a gate and locked thirty-six heifers off of water over a long weekend in August; the cattle suffered terribly and twelve of them died. High stock density grazing is to continuous grazing as a sports car is to a model T Ford; the sports car will take you faster and farther, but you had better pay attention to the road. When animals are concentrated, good things can happen to both the animals and the land but the situation must be monitored closely and continuously. The dangers to animals from poisonous plants, predation, parasitism, disease, and weather can all be reduced with good planning, but the manager must constantly observe what is taking place and make changes quickly when they are necessary.

- Needs of the overall operation. How to arrange to make the best use of time, labor, and available resources

A common reason people give for not using planned grazing is that it is too labor intensive. Having worked in both extensive and intensive operations, I can say without fear of contradiction that high stock density planned grazing can be a tremendous labor, time and energy saving practice.. I have spent many days riding big pastures when I felt lucky to see half the stock on any given day. Subdivide the pasture, and I could see every animal every day in a fraction of the time. On our Red River ranch, I could leave the house at daylight, move two sets of cows of about three-hundred head each, move a set of yearlings and a flock of fourteen-hundred sheep, and feed the guard dogs; I could see every animal on the place and get back, take a shower, drive thirty miles, and be on time for church. Ideally, I could do this, but there is a law somewhere that says that anything that is going to break, get sick, or crawl out will do it on Sunday morning.

After years of not sleeping the night before shipping day because I was worried about what sort of wreck could and probably would happen, once we and the cattle got used to the new program, we did not even saddle horses on shipping day. The cattle would be on daily moves and in the right place so that I could hook onto their mineral feeder, drag it into the lots, and every yearling would follow like a set of dogies following their milk bottles. Someone would follow along to shut the gate, and we would start loading trucks.

There can be an initial development period when fencing and water development require additional labor and expense, but I have seen many ranches that can do a very good job of planned grazing without building a foot of fence. It is true that someone must be available to check the water and move stock; if the plan calls for daily moves, someone must be there every day. If this is not possible, than a program with longer graze periods and lower stock density may be necessary; before giving up, however, realize that once facilities are in place and animals are trained, a twelve-year-old boy or girl on a four wheeler or a pony can move stock and check water just as well as a top hand. Temporary fence paddocks can be built ahead on the weekend so that moves become a matter of opening gates and checking water. Some may see it as a drawback, but stock gets very gentle, and the need for cowboying skills drops way off.

For most livestock breeding operations, winter forage costs are a major expense. With well-planned high stock density grazing, the use of hay or other stored feeds can be eliminated or greatly reduced. Rationing out dormant, warm-season grass or stockpiled, cool-season grass with temporary fence can be a high-paying job. Gathering stock for working or shipping becomes a nonchore, provided that plans have been made as to where the animals need to be at what dates. Events such as hunting seasons or crop harvests can be planned into the schedule to cause minimal disturbance and still meet the needs of the animals.

- Financial needs. How to do all of the above in the most financially beneficial manner

> *The secret to financial success is to do those things that make money and not do those things that cost money.*

That sounds trite, but it is true; the only way to generate wealth is to take in more than you spend. One of the strongest advantage points of planned grazing is that it allows management to be substituted for inputs of all kinds, thus saving money. Purchased or stored feed is a good example; a combination of correct stocking rate, good stocking mix, and high stock density reduces the need to use anything except grazed pasture. Productivity of pastures grazed at high stock density increases dramatically even without purchased fertilizer; weed pressure drops, especially in areas grazed with multiple species, so another input is eliminated. Some of the early installations built to facilitate high stock density were far too elaborate and expensive because people didn't realize how docile and easy to control animals become when they learn that moving is a good thing and they are moved regularly; short graze periods are good for the forage and the soil, but they also create content and easy-to-handle animals.

> *The one thing you can control absolutely is how much you spend.*
> *If you can't control spending, you need to re-plan.*

Each of these elements (human, land, animal, financial, whole) will have short-term goals (feed the cows today), midterm goals (provide forage for the entire season), and long-term goals (improve the forage base). Any goal must provide benefit to the element it addresses; it should also offer benefits to all parts of the operation, but at the very least, no goal should be achieved at the expense of permanently degrading other parts of the whole.

Once the goals are defined, it becomes possible to plan what needs to be done to realize each goal. When any action is considered, the deciding factor for acceptance or rejection should be whether the action will take you closer or farther from your goals. With constant change in a large number of factors (weather, markets, cost of inputs, etc.) plus the occasional disaster such as wildfire or disease, goals quickly become extremely difficult to manage effectively without a well-thought-out and monitored plan. It is not possible to control any of these factors, but it is very possible to reduce adverse effects through planning. A well-managed ranch in west Texas suffered repeated fires started by the railroad that bordered them until they built a long, narrow paddock along the railroad and reduced the amount of grass on it by grazing it continuously with bulls and saddle horses. This area was intentionally abused to spare the rest of the ranch; is this a violation of the edict to "do no harm" or just applied common sense?

One effect of the successful grazing plan must be to increase the health and stability of the soil and the forage that grows in it. A prime purpose of the plan should be to maximize the total benefits derived from the available resources; these benefits will include the growth of biological capital as well as the growth of fiscal capital. To be sustainable, an operation must be profitable, and no operation can be truly profitable without being sustainable. For any plan to be valid, it must begin with a clear vision of what is to be accomplished with the plan and there must be ways of determining its effectiveness..

To be valid, a grazing plan must advance the well-being of all parts of the whole operation.

Improving the health of the land

Manipulating the ecological processes mentioned earlier provides the means of improving the health of the land; monitoring changes in the conditions of these processes gives a method for determining from an ecological standpoint whether or not our management is effective. When the water cycle is effective, the mineral cycle is effective, and energy flow is high; long-term growing conditions will be good, and both forage productivity and stability will be good. By monitoring for changes in the factors that determine the health of the ecological processes, we can predict the effects of our management and make needed changes early. Any practice that reduces the effectiveness of the water cycle, decreases the amount of minerals in cycle or the rate of this cycling, or reduces the amount of energy flowing through the system will, over time, drive biological succession backward and will bring about a reduction in both production and stability.

The conditions of the ecological processes (water cycle, mineral cycle, energy flow, and biological succession) determine the health of the soil-plant-animal complex.

To improve the water cycle

- Maintain soil cover, either live plants or litter, to increase water infiltration and to decrease evaporation. Under good grazing management, bare ground will be reduced, as will capping of the soil. Uniformity of grazing promotes the health of both mature plants and seedlings.
- Increase soil organic matter content and thus soil tilth and soil life; keeping the soil covered is critical to promoting soil life. The cyclic growth and dieback of plant feeder roots is one of the best ways to increase soil organic matter. Well-managed rotational grazing promotes this process. Soil organisms are stimulated by the nutrient pulse that forage plants release into the rhizosphere when bitten and by the energy surge of manure, urine, and trampled plant material that comes with each graze period.
- Maintain enough plant canopies to break raindrops, trap snow, and moderate temperatures and wind effects. Use fire sparingly and only to treat (the rare) conditions that cannot be remedied with high stock density.

To improve rate and quantity of nutrient cycling

- Increase the number of species of plants and animals present; the mineral cycle is healthiest when all resources (mineral nutrients, moisture, sunlight, and space) are fully utilized but not overutilized by plants and the plants are fully utilized but not overutilized by animals. The pulsing action of animal impact with

high stock density grazing combined with a lessening of overgrazing promotes increases in the numbers of species present.
- Maintain a large percentage of the plant population in a vegetative state. Minerals locked away in senescent plant material are lost to the cycle until this material decays. As stock density increases, uniformity of graze improves so that by controlling recovery periods, the forage sward can be kept close to the desired physiological state. Use of multiple species of grazing animals increases the number of plant species utilized and improves both nutrient cycling and energy flow.
- Return ungrazed plants to the soil surface to decompose. Through trampling high stock density gets old material to the soil surface where decay is possible; in addition, the amount of forage refused and allowed to become senescent decreases under high stock density. Herd effect and dormant-season high stock density grazing can be used to break down old material that would otherwise be slow to recycle. Increase soil organic matter content and, thus, soil biological activity. Use grazing management o promote the cyclic growth and dieback of plant feeder roots to increase soil organic matter. Soil organisms are stimulated by the nutrient pulse forage that plants release into the rhizosphere when bitten and by the energy surge of manure, urine, and trampled plant material that comes with each graze period.
- Avoid any practice that decreases biodiversity or biological activity. Find an alternative solution to using any material that has a name ending in "cide" (death);

agriculture is supposed to be the art and science of promoting life. Especially damaging are herbicides, insecticides, tillage, and acid salt fertilizers.

Agriculture should be the art and science of promoting life so that we can harvest some of the surplus for our own use.

To increase biological energy flow

- Maintain green and growing plants in the area for as much of the year as is feasible to capture and convert as much solar energy to biological energy as possible. Mixtures of warm- and cool-season plants combined with control over when, for how long and to what extent an area is grazed, can greatly extend the growing season.
- Maintain a large percentage of the plants present in a vegetative (actively growing) state, as young tissue is both efficient in photosynthesis and nutritious. Avoid removing too large a percentage of the leaf area, because this slows plant recovery and decreases energy production. As stock density increases, uniformity of graze improves so that by controlling recovery periods, the forage sward can be maintained at approximately the desired age.
- Use mixtures of plants (warm- and cool-season grasses, legumes, forbs, and browse). Controlling stocking rate, stocking mix, timing of grazing, and stock density in ways dictated by the results of monitoring will allow complex mixtures to flourish.

To optimize level of biological succession

- Decide which stage of succession best fits your needs, and tailor management to suit that level while understanding that all parts of the soil-plant-animal complex are in constant and dynamic interaction. By manipulating water cycle, nutrient cycle, and energy flow, the manager determines the quality of long-term growing conditions. High succession requires good long-term growing conditions. All aspects of these processes must be good for succession to advance.
    - Example: Crop land is low on the successional ladder due to low biodiversity, low soil life caused by tillage, and low energy flow (short green season). Multiple-species–planted pasture is higher succession due to more biodiversity, less disruption of soil life, and more energy capture. Low succession does not necessarily mean low production; choose and manage for the levels of succession that best fit your overall goals.
- Monitor the results, and change management decisions whenever the results don't promote your landscape goals. Higher levels of succession may not always be best; the level of succession promoted depends on the goals.

Monitoring to ensure that the plan is providing the desired results must take place on both a short-term (was the correct amount of forage provided for today?) and a long-term (are favored species increasing?) basis, as all actions have both long- and short-term results. There are a number of good monitoring processes available, but Charley Orchard's Land EKG makes it simple to relate changes on the ground to changes in the ecological processes; check it at www.landekg.com.

In addition to monitoring the health of the ecological processes to understand how management is affecting the land, it is critical that three other areas be simultaneously scrutinized: animal performance, financial health, and the total well-being of the humans involved. Animal well-being is vital, and most problems of animal health and poor performance are preventable with planning and close observation. When we concentrate animals, we reduce their ability to select their diet and to move away from unhealthy or unpleasant conditions; a big part of the monitoring process is to make certain that all of the needs (nutritional, physical, and psychological) of the animals are being met all of the time. If animals are stressed in any way, performance will suffer, and disease will increase; job one is to keep the animals as happy as possible. Contrary to what has been reported in some quarters, it is not necessary to give up individual animal performance when changing to a high stock density program. If a program does not produce both higher stocking rates and better animal performance, the program is either poorly designed or poorly operated (or both). Analyze what is required to produce the results desired, observe what is actually taking place, and adjust management to make it happen. I once had a long and rather heated (on his part) discussion with a young man who had a brand new Master of Science degree in range science; he was adamant that individual animal performance always suffered under rotational grazing, because he had conducted a thorough study of the literature, and that was what nearly all researchers had found. I started trying to explain where most of the trials had gone wrong with long graze periods, moving animals by the calendar, stressful handling, and failing to monitor and adjust the programs to actual conditions, but he was not interested and left unconvinced.

If you understand what the animals need both physically and mentally, and you conscientiously provide it, the only limitations on performance are the genetic abilities of the animals.

Long-term financial health can be achieved only through production and marketing practices that are profitable and that can be maintained over time. Elaborate and expensive facilities are seldom necessary or desirable; the key to controlling animals is to make them content and comfortable in their surroundings. Monitoring for financial health should be a constant search for purchased inputs that can be eliminated and for ways to add value to production without increasing costs.

The challenge is not how to reduce the cost of a practice but rather how to eliminate the need for the practice.

Any expense that has to be repeated on a regular basis should be viewed as an opportunity to improve management; the question should not be how to do something cheaper but rather how to eliminate the need for the expense.

Ranching is a business and should be managed as such, but the purpose of the business should be to provide quality of life to the people involved.

We are constantly advised that we must treat ranching as a business—but what is the purpose of the business? You may have the best grass and cattle in the county and may make so much money that you have to lease land to stack it on, but if you run your kids off and work yourself to death in doing so, your business is a failure. Be on the lookout for things that can be changed that would make life better for all involved. If you don't like what you are doing, quit and do something else; life is too short to spend it doing something you don't like.

Regardless of how beneficial to the land and how profitable a program may be, it is not a successful program unless it provides quality of life for the people involved.

Plan What?

The grazing manager works with four separate plans: the grazing plan, the stock flow plan, the land plan, and the financial plan. The grazing plan specifies which animals are to be located where and for how long. The land plan is a map showing where animals are to be held, how they are to be watered, and how they will be moved. The stock flow plan gives expected and actual numbers of the various types of animals grazed both within and between years. The financial plan should account for all expected and actual income, and expense and be a prime source of the information used to improve the ranch business. All four of these plans should be under constant review and revision, as the real world is constantly changing.

The Land Plan

The purpose of land planning is to provide the physical attributes (subdivisions, water points, shelter, etc.) required so that animals can be positioned in time and space in the most beneficial manner that is feasible, given the constraints of available labor, time, and money. The grazing plan details how these features will be utilized; it specifies how many animals of which types will be present at any time on all areas and how long they will be there. To make these assignments, it is necessary to know both forage demand (number and weight of animals by class and time that animals are to be present) and forage availability (forage growth rate, area of forage, and starting volume of forage). The grazing plan must contain reasonable estimates of the amount of forage available in each subdivision so amount of forage per unit of land must be considered when developing the land plan. If the land in question is fairly uniform in its ability to produce forage, this becomes largely a matter of dividing acreage by the number of paddocks planned. If there are large differences in forage production, the forage production of each paddock must be rated as a percentage of the best paddock; this can be done by clipping and weighing forage, but most people will have a pretty good idea from experience whether the back forty has half as much or 75 percent as much forage per acre as the front forty. Regardless of how these assessments are made, it is critical that they be checked and verified by experience when grazing starts.

If control of the animals is to be done with fencing, decisions must be made in two broad areas: (1) how many paddocks of what size and location and (2) how the animals are to be watered while in these paddocks.

Factors affecting the number, size, and construction of paddocks

The number and size of paddocks and how they are constructed will be determined by the following:
- Number of herds to be used
- Size of herds
- Number of different types of animals to be run
- Length of graze periods
- Length of recovery periods
- Amount of forage per unit of land
- Number of different types of areas that need similar management

- Number of herds to be used

How many groups of animals are to be kept separate? Why are there so many groups? Which groups can be combined either permanently or temporarily?

Ranchers seem to have a passion for splitting up their animals that can only be satisfied when every available subdivision has its own permanent resident herd. I went to a cow-calf outfit several years ago that had twenty almost equally sized, permanently fenced pastures; all but one of these subdivisions had cattle in them. The day I was there, the cow boss had the hands riding out pairs from a set of calved-out cows; he was sorting the cows with bull calves into the one empty paddock and away from the cows with heifer calves so that "We can make those bull calves grow faster by giving their mommas a little extra." There are always reasons for grouping animals: spring calving vs. fall calving, stockers vs. cows, cows to be bred to bull A vs. cows to be bred to bull B. *Most of these reasons cannot contribute as much value as could be gained by having more paddocks per herd*. For certain, you don't need to run dry spring calving cows with wet fall calving cows—but the question is, why do you have cows calving in both the spring and the fall? Pick the season that works best for you in your country, and get all of your ladies on the same schedule. I have heard lots of reasons for having a dual calving season: double use of the bull battery, more even cash flow by selling twice a year, and so on. Perhaps the weakest excuse is so that a cow that comes up open in one season can calve the next season and not miss a full year.

Even if you have to breed a set of cows to a particular bull, you can split those cows out for a 45–60-day breeding season and then put them back with the herd. For years, I ran two herds of spring calving cows, because if we came to the pens to work calves with more than about 350 cows, it made for too long of a day and stressed both man and beast. I mentioned this to Stan Parsons one day, and he suggested that I gate cut the herd into two groups without attempting to pair up the calves, and turn the two groups out on either side of a single strand of electric fence wire raised just enough that a calf could walk under without getting bitten. Overnight, the two groups mothered up so that we could work one group one day and the other group the next day. Fewer herds mean fewer paddocks needed for good grazing management and less time and labor per head of stock.

The cheapest way to increase the number of paddocks per herd is to reduce the number of herds.

- Size of herds

    I know people who are running cows, weanlings, yearlings, and even grass fattening animals all together in one herd and, because of top-flight management, are doing a good job; however, most of us will probably be better off separating the breeding animals from, the stocker animals, and fattening animals. If this is done, the size of the herds becomes an arithmetic exercise. How many animal units can you run, and how will they be split up? A common excuse for small herds is weak water: "We can't put over fifty cows in that pasture, because that's all the well will water." If a well can water fifty cows for a year, that is 18,250 cow days of water; this means that if storage is provided, it will water a thousand cows for eighteen days. The amount of water storage needed will depend on how often the animals use the water. Storage is not cheap, but the benefits of being able to increase stock density are huge. It is not necessary that all animals be able to drink at the same time, but it is critical that the rate of water delivery to the drink tank be fast enough that animals cannot drink the tank dry; if this happens, the animals will go into panic mode, stress the entire herd badly, and probably tear up something. The herd is the basic social structure of all domestic livestock; being in large herds is not stressful to these animals, but our mismanagement of large herds can be very stressful. Gates in the wrong place, gates or lanes that are too narrow, or gates that have mudholes in them can cause a simple move to the next paddock to become a turmoil of calves pushed through fences, cows running back to find lost calves, wasted effort, time lost from grazing, and language you wish your kids hadn't heard. As the size of the herd increases, the time, equipment cost, and labor per head decreases—provided that the facilities are adequate for the large herds.

- Number of different types of animals to be run

Although different types of animals can be run together, I don't recommend that people attempt this until both they and their animals are experienced with high stock density. The main reason for separating different classes of the same species of animal is the differences in their nutritional demands. Mature breeding females have a high demand for energy and protein starting shortly before they give birth; this demand is highest in the weeks following birth, as the animals must repair the ravages of birth, start lactation, and get ready to rebreed. Nature fulfills these needs by having wild, grazing animals wait to give birth until several weeks after the new crop of forage has significant growth. This is the cheapest and lowest labor method that we can use to meet the needs of our breeding females. The nutritional demand for females is in direct proportion to the amount of milk they are producing; when the young are weaned, the demand drops dramatically and remains low until close to the time for new young to be born. Young growing animals and animals being fattened for slaughter have continuous high nutritional demands. Most people will find it advantageous to separate them from the cowherd in order to make use of the dry cows' ability to utilize lower-quality forage.

Sheep and goats have different fencing and management requirements, and most people will find it easier to run them as a separate herd. There are valid reasons to have multiple herds, but most of us have more herds than valid reasons; spend some time trying to combine some herds before starting the planning process.

- Length of graze periods

Recognition of the benefits of using short graze periods has been slow to develop; some of the early work done with rotational grazing seemed to suggest no range improvement benefit in having more than eight to ten paddocks per herd. Many of these early studies were of short duration, however, and set up on a calendar basis and few even attempted to measure animal performance. When private individuals began to experiment with rotational grazing, people began to use shorter graze periods and more paddocks per herd to better control the age of forage presented to the animals. The age control and other factors such as competition did result in better animal performance, and the higher stock density (as a result of the concentration of animals on smaller areas) produced healthier soil and forage. The value of short graze periods has been demonstrated by the results seen from strip or mob grazing, in which animals are moved multiple times a day. If only permanent fencing is used, the number of paddocks available in combination with the length of recovery periods required will determine how often animals can be moved; if temporary fencing or a combination of permanent and temporary fencing is used, the frequency of moves will depend on when labor is available and on the cost-to-benefit ratio of using shorter graze periods. Most commercial grazing operations can benefit from using one-day graze periods. By subdividing with temporary fence, dairy cattle might be provided with a fresh break of forage after each milking, and animals under the care of a full-time herder might be moved eight or more times per day. Length of graze periods is important to animal performance and to efficiency of the grazing process, but the length of recovery periods is critical to the whole operation and must not be shortened to reduce the length of graze periods. To reduce the length of graze periods, the number of subdivisions per herd should be increased.

- Length of recovery periods

In areas where growing conditions are good over long periods of time, forage will grow fast enough to be grazed multiple times per season; in these areas, grazing periods can be kept short and recovery periods correct with a relatively small number of permanent paddocks. Example: Low seral forages (Bermuda, fescue, clover, etc.) with good growing conditions can recover in twenty-four to twenty-eight days. If twenty-five paddocks per herd are available, the animals can be on one-day graze periods until growth begins to slow due to moisture or heat. At this point, extra time can be gained by keeping the animals longer in strong paddocks; this could be done by using multiple-day graze periods, but if feasible, subdividing the paddocks with temporary fence and staying on one-day graze periods will yield better results.

Higher seral plants such as big bluestem, Indian grass, or eastern gamagrass require longer recovery periods, so thirty-five to forty paddocks per herd would be a better fit. The number of permanent paddocks planned becomes a tradeoff between (1) the flexibility but higher labor demands of temporary fencing and (2) the lower labor but higher initial cost and reduced flexibility of permanent paddocks. Whatever their number, permanent paddocks should be planned with ease of subdivision in mind; it is possible to do an excellent job of grazing with all temporary fences except for perimeter fences. There is little or no benefit in coming back to graze dormant forage that has already been grazed; much better animal performance and forage utilization will come from taking everything that is to be consumed from dormant forage in one graze period.

- Amount of forage per unit of land

    When carrying capacities are high, the expense of water development and fencing is easier to justify, as it can be spread over a lot of animals; when stocking rates are low, overhead and infrastructure costs must be limited. In areas with very low stocking rates, the only feasible subdivision may be to turn water points on and off and loose herd stock from one area to the next. Use a sharp pencil and figure the costs carefully, but don't automatically rule out some subdivision on low stocking rate areas; in these areas, poor animal distribution can create large areas of underutilized forage that, with a reasonable amount of expense, could be brought into use to increase stocking rate. Where carrying capacities are high, it is easy to spend more money than necessary on facilities; work hard at reducing the number of herds being run before going on a development binge. Try to lay out paddocks so that they can be subdivided easily; it is often beneficial to lay out large, permanent, fenced paddocks and then subdivide them with temporary electric fence at least until you are comfortable with the degree of subdivision.

- Number of different types of areas that need similar management

It is possible to use temporary fencing to split off areas within a paddock that need to be treated differently from the rest of the paddock, but planning and implementation are much simpler if paddocks are laid out so that the entire enclosed area benefits from the same treatment.

Common examples of such divisions would be:
- tame grass
- native grass
- cropland
- timber
- riparian areas

This is a good time to decide whether changes should be made in land use; is it time to take the fence down from around Grandpas' cotton patch and take it back to grass?

The availability of very good and inexpensive electric fencing has made high stock density grazing feasible in most areas. Some very good operators use only temporary electric fencing, but most use a combination of permanent physical barrier fences and temporary electric fences. If graze periods must be long (four days or more) because labor is not available to move animals, animal pressure on fences increases greatly. Animals being moved using short graze periods are very easy to control because they are content. Very short graze periods involving multiple moves per day are usually best handled with either herding or temporary fencing; one of the factors to consider when planning permanent fencing is how easily the permanent paddocks can be subdivided with temporary electric fence. Square paddocks require the least amount of perimeter fence per unit of area enclosed but require more feet of temporary fence and time to subdivide.

Land planning can provide the tools needed so that one, two, or three of these factors (time, area, and number of animals) can be manipulated for management purposes. By planning and managing the length of periods in which areas are exposed to grazing and the lengths of periods they are protected from grazing, a good manager can create the conditions favorable to desired plant species, bringing about an increase in these species. The same tools can be used to improve animal performance in both the short and long term and to control pests ranging from weeds to parasites. Grazing management can be used to improve conditions for all aspects of the operation (forage, soil, animals, finances, and quality of life) for the operators, but this is possible only if the position of animals can be controlled at all times. This can be and is accomplished with herding, but most people will find that the use of subdivision with fencing better suits their needs and abilities.

The land plan map shows where and how animals are to be located; these subdivisions (paddocks, pastures, whatever's) may be temporary and flexible, as when herding or temporary fence is used, or they may be defined by permanent fencing. Regardless of whether it is temporary or permanent, all of the area within a subdivision will receive the same treatment with regard to length of graze period and degree of forage utilization; in most cases, the length of recovery periods will also be the same. As different types of forage require different management techniques, all of the area in a subdivision should have the same type of vegetation; native grass should be separated from tame grass, and both should be separated from cultivated crops. If an area of native grass is enclosed with an area of tame grass, one or the other will not receive the management it needs to thrive. Virtually all tame grass (the so-called improved grasses) are low seral (adapted to poor growing conditions) and able to withstand abusive grazing. That adaptation plus the ability to make use of nitrogen fertilizer is why they are termed "improved." It seems that we would rather search out and breed plants that can withstand poor management than improve our management. The quality perennial native grasses require longer recovery periods and need more leaf residual than do plants such as Bermuda grass and endophyte-infected tall fescue. If a paddock with both high-quality native grasses and low seral plants is grazed in a manner to maximize the contribution of the low seral plants, the native grasses will be killed out; this will happen even faster with high stock density grazing than with continuous grazing, as few of the native plants will ever become old, tough, and woody enough to be protected from grazing. If the management of a mixed-type paddock is grazed to favor the native grasses, they will predominate over time; long recovery periods will mean that the low seral plants are always grazed when in phase three, and their value as forage will be reduced. Management will be much easier if forage types are separated. Paddocks should also be defined, when feasible, by topography; steep slopes will be underutilized if they are enclosed with level ground, and riparian areas will be overutilized unless access is limited.

Benefits of Planned Land Subdivision

- Increase the quantity of usable forage by increasing the efficiency of grazing. This can result in an increase of 40 percent over continuously grazed areas.
- Improve the function of the ecological processes through stock density and time control, and thus create a more productive and stable grazingland.
- Use recovery periods and degree of utilization to change the seral level of the vegetation of an area.
- Provide the means to use animals to modify existing vegetation.
- Increase animal performance by presenting a clean, dense sward of forage at the proper state of maturity.
- Improve animal health by reducing parasite and disease infections.
- Reduce the need for supplemental feed by providing quality forage over a larger part of the year and by rationing out dormant forage.
- Increase animal performance by reducing the stress of handling on animals.
- Reduce labor costs.

- Create a more flexible grazing program.

Process of Land Planning

Questions to be answered before any fence is built

- Which areas need to be fenced off from neighboring areas?
    - All of the area within a subdivision will receive the same treatment, so only like areas or areas to be converted should be included in a subdivision.
- Is the land leased or owned? If leased, have agreements been made as to who owns what at the end of the lease?
- What level of stock density is planned for both now and the future?
- Which species are to be grazed, and in what combination? Is this apt to change in the future?
    - Different species have different fencing, watering, and management requirements that should be considered before construction begins.
- Which existing water points are suitable for use in a high stock density paddock?
- What are the water development options?
- How many herds and how many paddocks per herd? Is the use of temporary fencing to be used to subdivide permanent paddocks?
- What size herds?
- In the final plan, will every paddock have its own water point?
- Will the improvements be built over time or all at once?

- If building over time, which facilities should be built first?
- In the final plan, will paddocks be close to the same size, or will some be planned to be larger or smaller?
- Are lanes planned for stock movement?

Land Planning Tips

Have on hand as much information as possible about the area to be planned. Plot on the map the areas that will receive different treatment: cultivated land, native grass, tame pasture, timber, and the like. This is a good time to decide whether these areas should remain in their present state. Is it time to take the fence down from around old cultivated ground and take it back to grass? Think about the topics of all of the stated questions with the goal of maintaining as much flexibility as possible. Try to reduce the number of herds run to as small a number as possible; often, it is possible for established ranches to convert to high stock density grazing simply by combining herds.

Plan the entire ranch just as though you were going to build all improvements tomorrow.

It is much easier to change a line on a map than to move a buried water line. Begin construction with those features that offer the best return per dollar spent; often the best return will involve splitting the largest paddocks before further subdividing the smaller paddocks.

Major consideration should be given to water quantity, quality, reliability, and cost of development. As stock density is increased, the demands on water points become much greater. An individual water point in each paddock is the ideal situation, but this is seldom feasible. Planning to make the best possible use of the water available should be a main priority.

All forage within a paddock will receive the same grazing management, so forages with different seral levels (and thus requiring different lengths of recovery periods) should not be combined within a paddock. Different paddocks within a cell may and sometimes should have different landscape descriptions; thus, they require different lengths of recovery periods. Example: Bermuda grass paddocks within a cell of native grass paddocks might be grazed on a shorter recovery schedule than the native grass paddocks to increase animal performance.

The amount of land within a cell will be determined by the size of the herd or herds using the cell and the carrying capacity of the land. Plan to develop the optimum number of paddocks for each cell. This number will vary according to the range in length of both graze and recovery periods needed to promote the desired landscape description and good animal performance. A reasonable starting point for number of paddocks per herd is one that provides enough paddocks for the animals to be on one-day graze periods when growing conditions are best and recovery periods are shortest. This is feasible, of course, only if someone is available every day to move the animals. The number of paddocks used will always be a compromise between costs and results; a large number reduces labor but also reduces flexibility and increases construction costs.

If multiple species are to be grazed, consider special needs, such as having kidding paddocks large enough that nannies do not need to be moved during kidding. The recovery time required for browse is much longer than for forages and must be considered when planning construction. Leaving large amounts of functioning leaf on the browse plants at the end of each graze period can reduce the need for longer recovery periods. It may be possible to lower fencing costs for sheep, goats, deer, and the like, and still meet their management needs by constructing a smaller number of goat tight paddocks that are subdivided with single-wire electric fence so that cattle can be rotated through the goat tight paddocks on a shorter graze period.

Consideration of animal behavior and to physical features of the land is vital when designing paddock layouts. Thought given beforehand to ease of animal flow and prevention of problems with steep slopes and boggy ground will pay big dividends later. Whenever possible, limit the amount and number of areas that receive prolonged or repeated use. Water lanes and excessively large cell centers are sources of disease and parasite infection and are soon degraded by trampling and erosion.

Animal performance begins to fall off when cattle have to travel over seven-hundred feet for water, and the closer to square a paddock is, the better the efficiency of grazing. If a combination of permanent and temporary fencing is used, elongated paddocks require less temporary fence for subdivision.

Do not become locked into making use of existing facilities. Only those features that cannot feasibly be changed should be considered permanent. Often, fences were built for reasons that are no longer valid. Plan the area the way you feel it should be done without regard for present fences. If present fences work, that is fine, but don't create a flawed plan to save a fence. It may be helpful to plan to leave some present fences in place until you are ready to build the part of the development that they affect.

Realize that as your new management kicks in, the number of stock will rise. Water, especially, should be planned to accommodate the larger numbers. Water that was adequate when a pasture was set stocked with fifty cows may be totally inadequate to water a combined herd of three hundred.

After the plan is mapped, move a "paper herd" through the rotation to see if stock will flow or if there are areas where the planned movement will conflict with the normal behavior of the stock. Is the plan flexible with regard to stock movement? Be careful to not create a situation where you can only go from paddock 10 to paddock 11. Does the plan allow for paddocks to be dropped from the rotation without creating long or awkward moves?

Take the map to the field and check to see if your memory was correct regarding the topography and the location of features such as creek crossings and wet spots. Now is the time to prevent future problems such as mudholes in gates and places where baby calves can get cut off.

Spend whatever amount of time is necessary to arrive at a plan that you like, and then put it away for a few days and see if it still looks as good when you come back to it. It's a lot easier to build fence and dig water line on paper than on the ground.

Stock Flow Plan

In its simplest form, a stock flow plan is simply a day-to-day inventory by class and type of the livestock present; the anticipated inventory should be projected into the future as the forage demand estimate for use in the grazing plan. If significant wildlife is present, an estimate of their numbers and forage demand should also be included. One of the most valuable benefits of keeping good stock records is that it allows the manager to forecast forage demand with a high degree of accuracy. If the stock flow chart is set up as a spreadsheet with the various classes in permanent headings under months, it is simple and quick to make changes as animals change class or location. Forage demand for various classes can be varied as the season progresses to reflect changes in weight as animals grow. The Animal Unit factor used in the following spreadsheet can also be manipulated to regulate the amount and quality of forage presented to various classes. Example: If the goal is to fatten a group of heavy slaughter cattle as quickly as possible, their AU factor might be increased to 1.25 or even 1.5 instead of the 1.00 that their thousand-pound average weight would otherwise dictate; this has the effect of increasing the amount of forage put on offer so that selection is better, and thus, weight gain is faster. Forage demand is a function of weight, but it is also a function of animal physiology; young growing animals, lactating animals, and fattening animals require more nutrition than do adult animals that need only to maintain their already-grown bodies.

The stock flow chart is also very useful when combined with a forage growth chart for determining realistic overall stocking rates as well as stocking rates by class of animal.

One of the most beneficial pieces of information to the grazing manager is when and in what quantities forage grows on a particular area. This can vary from something like 150 pounds of forage an acre that grows in the sixty days immediately following the monsoon rains and before a killing frost, to 80 times or more that much forage growing over the entire year. Learn the normal growth patterns of your operation; they may or may not be typical of your area, depending on your management. Use this information along with a "what if" stock flow chart to arrive at a stocking rate and stocking mix that make sense for your operation.

There are some software programs that can be of great help in predicting both forage growth and forage demand, but don't make the mistake of allowing the computer to do your thinking; computer programs are just another way to plan, and like all plans, they must be monitored and adjusted and sometimes thrown out. Research the needed information, and do the best job possible of setting stocking rates, but remember that your estimates are almost certain to be off the mark. The only way to be certain that stocking rates are realistic is to be constantly monitoring forage growth and disappearance and plotting these values against known future demands. Use the projected animal numbers from the stock flow plan to determine future forage demand, and correlate this figure with present forage production and usage; a forage reserve for use during slow growth or during the dormant season can be built only when plants are actively growing. If you are fully stocked during the spring flush, you had best have plans for those animals to be somewhere else in January. Stocking rates that are too high are responsible for much of the lack of profitability that plagues many ranches. It is possible in many places to increase stocking rates by adding outside feed and supplements, but this increased production normally comes at the expense of dramatically reduced profit margins, long-term reduced forage production and reliability, and increased financial risk.

You need to decide whether you are running a ranch or a feedlot.

STOCK FLOW CHART

January 2011

Class	Number	AU factor	AU
Dry cows	100	1	100
Wet cows	0	1.2	0
H1 heifers	20	0.75	15
H2 heifers	25	1	25
Weaned calves	95	0.5	47.5
Yearlings home	79	0.75	59.25
Yearlings purchased	80	0.75	60
Bulls	4	1.5	6
Horses	2	2	4
Ewes	500	0.125	62.5
Lambs	650	0.07	45.5
Bucks	12	0.2	2.4
Deer	30	0.125	3.75
Total AU for month			430.9

February

Class	Number	AU factor	AU
Dry cows	85	1	85
Wet cows	0	1.2	0
H1 heifers	20	0.75	15
H2 heifers	25	1	25
Weaned calves	95	0.5	47.5
Yearlings home	69	0.75	51.75
Yearlings purchased	75	0.75	56.25
Bulls	4	1.5	6
Horses	2	2	4
Ewes	500	0.125	62.5
Lambs	650	0.07	45.5
Bucks	12	0.2	2.4
Deer	30	0.125	3.75
Culls	15	1	15
Total AU for month			419.65

The stock flow plan is useful as an inventory, but its real value is as a planning tool. The amount of forage available varies according to the season of the year; one of the best profit-generating practices available to ranchers is to match the demand for forage with the supply of forage. This match must, of course, address quantity of forage but should also address quality. Most arid and semiarid ranches will grow their forage in a narrow time frame and spend the rest of the year rationing out this supply to their animals. The better the job of rationing, the better-quality forage the animals receive, and the better the animal performance. Even with excellent grazing management, the quality of forage goes down as it ages but this loss of quality will be much less with planned grazing as compared to continuous grazing. In more humid areas with a longer growing season, forage quality can be high for longer periods of time, but the principles are the same. By planning the stock flow—which animals will be present during which periods of the year—it is possible to increase the value of the forage harvested. In general, animals with high nutritional demands should make up a larger part of the stocking rate when forage quality is high than when forage quality is low. It is possible to run stocker animals or milking cows on frosted forage in January, but acceptable performance will likely require either putting large amounts of forage on offer or supplementing the ration. Run inventories on the normal forage by both quantity and quality by month, and use this information to plan out a stock flow that fits the situation; as with all plans, monitor the results closely, and be ready to change the plan when necessary.

Grazing Plan

Why plan grazing?

The practice of livestock grazing, especially on public lands, has a history of controversy, with some people contending that all grazing is destructive. It might be valuable to look at the history of public land grazing to understand how and why public perceptions of grazing were formed. In the late 1800s and early 1900s, livestock grazing on public lands in the western United States exploded; the slaughter of the bison herds and the pacification of the plains Indian tribes opened millions of acres to livestock grazing for the first time. The first cattlemen to exploit this resource did very well financially, and their success attracted investor money from all over the world. Livestock syndicates were formed to buy cattle and sheep to be run on the "free range," and investors made excellent returns on their investments for a number of years. As time went on, human nature intervened, and the maxim "What belongs to all will be cared for by none" was proven once again. More and more stock was funneled on to the public lands until most were stocked well beyond their sustainable carrying capacity. Practically all of the public land in the West was damaged to some extent, and many fragile areas went from productive grassland to desert in a few years. There is no doubt that tremendous damage was done and that it was done by grazing, but at the time, a common attitude was that removal of the grass would make it easier for farmers to plow the land for crops and "improve" it.

Much of the early work in range management in the United States came about as the result of this damage done to public lands. This early work centered, rightfully so, on reversing the damage to the forage and to the soil. It became obvious that the ranges were overstocked when hundreds of thousands of cattle starved and died in the series of droughts and blizzards that swept the U.S. plains in the late 1880s. It appeared obvious that the damage was caused by too many cattle, so common sense suggested that the way to correct the damage would be to reduce cattle numbers. This seemingly logical but only partially correct conclusion would hold up a deeper understanding of range ecology for many years. Only fairly recently have most range scientists agreed that damage to grassland cannot be halted solely by reducing stocking rate. There definitely were too many cattle, and the ranges were badly overstocked, but the damage to the grasslands involved more than just livestock numbers. Overstocking is damaging to grasslands, but overstocking is only part of the problem and can be reduced without halting the damage. Factors other than stocking rates contributed to the damage; chief among these factors was the fact that grazing animals were continuously present, preventing forage plants from recovering from the defoliation. Too frequent defoliation is the cause of plants being over grazed. Reducing stocking rate on a continuously grazed range will not halt overgrazing; it reduces the number of plants being overgrazed but increases the number of plants damaged through underutilization. This statement can be verified by observing any continuously grazed grassland. In an area that is continuously stocked at anything like a reasonable rate, by the end of the growing season, many of the plants are in an unhealthy state. Plants that were grazed early in the season regrow, and because their foliage is fresh and tender, they are grazed again before they have time to fully recover from the first defoliation; depending on the situation, this may occur multiple times within a season. Plants that were not grazed in the early part of the season (there will be many of these in a reasonably stocked area) have a lot of volume, but, because they have toughened with age, they are passed over by the cattle in favor of the tender regrowth as long as it is available. By the end of the season, many plants are in an unhealthy state; some of the plants have been damaged by overgrazing and some by underutilization. Ungrazed plants store a lot of energy in crowns and roots but also shade their own growing points, restricting growth. If this lack of defoliation continues for several seasons, these plants begin to die.

The situation created by continuous grazing can be deceiving; the overgrazed plants may appear to be green and healthy as long as growing conditions are good, and the ungrazed "wolf plants" give the impression that a lot of forage is present and that stocking rates are not too high. However, the damage done to grassland by continuous grazing is demonstrated by the fact that it is frequently possible to sustainably increase the carrying capacity of a range that is being grazed continuously at recommended stocking rates by 40 to 50 percent and at the same time increase the health of the range by changing the method of grazing. Concentrating the livestock and moving them through the available area, grazing one parcel at a time with sufficient recovery time between graze periods, can stop the overgrazing and increase the uniformity of utilization; the overall effect is to increase the amount and quality of forage grown and the efficiency of its utilization. The concentration can be achieved through fencing or herding; the objective is to control what and to what extent the animals graze at every point in time.

> *Proper grazing builds healthy grasslands.*
> *Abusive grazing destroys grasslands.*

All of the great natural grasslands of the world evolved with grazing animals in a synergistic relationship, and the duplication of this relationship should be the goal of any grazing management program. The key word is synergistic, meaning mutually beneficial; to be successful in the long term, a relationship must be beneficial to all parties in the relationship. This is true in business, in human relationships, and in ecological relationships. The great natural grasslands, regardless of where they are located, were formed with man playing little or no role other than in some spots being another pack hunting animal; the grasslands were formed long before livestock was domesticated. The formation of grasslands was a natural process, but nature will not restore the grasslands that have been degraded; man has become too numerous, too powerful, and too ready to intervene. If our grasslands are to be made healthier, it will only come about through intelligent management by humans. Much has been written on the subject, but good grazing management boils down to managing grazing livestock so that the needs of the animals, the forage, and the operation as a whole are all met. If any part of the whole under management is degraded, sooner or later the program will fail. It is possible to take short-term financial gain at the expense of the health of soil and forage and to make short-term improvements to soil and forage at the expense of profitability. Neither option is sustainable. Perhaps Alan Savory's greatest accomplishment was in forcing us to realize that we must manage to benefit all aspects of the operation; our decisions must be sound ecologically, financially, and sociologically. To succeed, we must manage the whole.

To be successful, a relationship must be beneficial to all members of the relationship.

When man came on the scene, he began to try to protect his domesticated animals from the predators, first by herding, then by fencing, and then by killing off predators. The more successful this effort, the more his livestock lost the herding instinct and spread out rather than staying in compact herds. Today, most livestock, even sheep and goats, have weak herding instincts and, unless they are confined or herded, spend much of their time dispersed in small groups. These individuals and small groups tend to locate on favored areas, grazing and regrazing some plants while other plants that have been fouled with dung or simply not bitten during the abundance of the spring flush become overly mature and senescent. Forage plants, those plants that evolved with grazing animals, need the stimulation of occasional defoliation to thrive. It is typical of areas grazed continuously at low stock density (which is the norm in most of the world) to simultaneously have some plants suffering from overgrazing and others from underutilization. It is also common to have whole areas receive little or no grazing pressure, simply because they are out of the way or harder to graze; these areas tend to be bypassed when forage is abundant, thus, their plants become tough and unpalatable and are grazed little even when forage is short.

There are many benefits to mimicking nature's grazing management, but even if we consider only the effects of reducing the overgrazing and underuse of forage plants and of reducing the forage lost to spoilage when animals roam back and forth over the same area, the benefits are huge. If an area is being grazed continuously at a reasonable stocking rate, it is usually possible to increase the stocking rate by 40–50 percent simply by increasing the efficiency of grazing. This is accomplished by limiting the movement of animals so that the area exposed to grazing at any one time is controlled, allowing the lengths and timing of both grazing and recovery periods to be set at durations favorable to both plant growth and animal performance. This plus the reduction in forage spoiled by trampling and dunging under continuous grazing simultaneously brings about an increase in the amount of forage that actually gets into the livestock and improves the nutritional content of the forage consumed.

These benefits are real and quite large, but even more significant benefits show up as the effects of high stock density bring improvements to the ecological processes.

Much of conventional management is directed toward fighting against what we don't want. We expend time and money fighting problems (weeds, diseases, and pests), even though these are really not problems at all but rather are symptoms of weaknesses in our management. Healthy, diverse grasslands seldom have weed problems unless the weed spray salesman convinces you that dotted gay feather, filaree, prairie mimosa, partridge pea, and four-wing saltbush are weeds.

Weeds become a problem when the growing conditions in an area favor the growth of weeds (plants adapted to poor growing conditions) rather than the growth of desired plants. Thus, the way to replace weeds with desirable plants is to change management so that growing conditions favor these desirable plants. The most common poor growing condition is too frequent defoliation of forages; too severe defoliation is usually second.

Animals in a low-stress situation on a correct diet with a high plane of nutrition and moving regularly to clean pasture suffer little from disease and parasites. Parasites and diseases can be controlled more effectively and more economically with management than with poisons and antibiotics. We can reduce parasite loads by encouraging biological activity in and on the soil. Horn fly eggs don't turn into flies if the dung they are laid in is buried or desiccated by dung beetles, and the ones that do hatch don't cause problems if they if they are eaten by spiders or predatory wasps. If the cattle have moved to more distant paddocks by the time the horn fly eggs hatch, many of the new flies will not find the herd and die. Internal parasites can be controlled by grazing multiple species, moving the animals regularly to clean pasture, culling animals with poor resistance, and leaving high forage residual in the grazed area. Disease is best controlled by reducing stress of all kinds:

physical, nutritional, or physiological. Our job as stockmen should be to keep our animals just as happy as is possible. Elsie, the contented cow, really does give more milk, and animals that are moved regularly to clean pasture of the proper kind, age, and density are in cow heaven.

Parasite and disease outbreaks are caused by faults in management.

The need for nutritional supplements can be reduced or even eliminated if production schedules are timed to make the periods of highest nutritional demand to coincide with the periods of peak forage production. The amount of forage grown can be increased by controlling the degree and frequency of defoliation, providing adequate recovery periods between graze periods, and using grazing as a tool to improve soil health. The percentage of forage that actually benefits the animals can be greatly increased by using short graze periods; short graze periods are beneficial to ration out forage, to improve animal performance by maintaining a uniform diet, and to increase the efficiency of grazing. The effectiveness of grazing will be increased if the quality and amount of forage presented closely matches the nutritional needs of the class of animals being grazed.

Planned grazing with good time control is the most powerful tool I know of to achieve all of these benefits. Planned grazing requires thought and planning but offers tremendous benefits in reduced labor and purchased inputs as well as in increased productivity and profitability. A planned grazing operation usually requires at least some expense for fencing and water development, but over the years, I have seen many ranches that could put good high stock density planned grazing in place without building a foot of fence. The cheapest way to increase paddocks per herd is to reduce the number of herds.

To stop overgrazing, the animals should be removed from grazed areas before they begin to graze the regrowth of previously grazed plants, and they should not return to that area until the plants have fully recovered. The concentration of animals will cause a higher percentage of the plants in the area to be bitten so that the number of slow-growing, underutilized plants is reduced, and more of the less favored plants are utilized. In continuously grazed pastures, the favored forage plants carry a disproportionate amount of the forage demand and will, over time, be damaged or killed. The spreading of the forage demand over all of the forage plants (plus some of the non-forage plants) by increasing stock density and reducing the length of the graze periods allows the favored plants to become more productive. The health of these favored plants will be further improved if the recovery periods between grazing periods are timed to meet their needs. The amount of time that forage plants need to recovery from defoliation depends on the quality of the growing conditions, the degree of defoliation, and the growth habits of the defoliated plant. Many of the most productive native grass plants are slow maturing and produce much of their foliage in upright structures that are easy for animals to graze; they stay in the phase 2 growth stage longer than do most forage plants. Thus, while providing high-quality forage for long periods, they also require longer periods of time to recover from defoliation. Although a plant like Bermuda grass might be fully recovered in twenty-eight days, under the same growing conditions and percentage of defoliation, big bluestem might not be fully recovered until fifty days or more have passed.

Continuously grazed ranges, even if not overstocked, will always be degraded to some extent, because both too frequent defoliation and lack of defoliation are harmful to forage plants. A healthy grassland is extremely resilient and resistant to change; couple this fact with the wide range of year-to-year differences in growing conditions that are normal in the climatic regions where grasslands form, and it is becomes very difficult to detect gradual changes in range health due to management. Even trained research scientists fall victim to trying to extrapolate data from short-term projects into valid, long-term management programs. Several years ago, I spoke to a group of ranchers in the short-grass country of western Kansas. I went through some of the advantages that I personally and some of my clients had obtained by practicing planned high stock density grazing in areas similar to the one they were ranching. My observations were not based on a short-term research project but rather on a real need to know combined with forty-some years of reasonably well-educated experience. When I finished my presentation, a range "professional" who had just completed a three-year grazing research project rose to tell the group, "If you do what Mr. Davis recommends, all you will do is increase your costs and reduce your production. This country is so resistant to damage from grazing that the only reasonable way to graze it is continuously at a high stocking rate." That quote is as close to exact as

I can remember it, and I remember it quite well. In the three-year study, the treatment that made the most money was continuous grazing at a high stocking rate, so this was the management that he recommended the ranchers follow. It is not criminal to be stupid, but it should be criminal to teach stupidity. I don't tell this story to throw rocks at academia, though I occasionally run into academics educated well beyond their intellectual capacity, I tell it in an attempt to get people to question, to observe, and to arrive at their own conclusions.

In any discussion of grazing management, the following statement will come up: "Research shows that rotational grazing improves gain per acre and decreases gain per head while raising costs." This statement is as valid as, "You can't make a success of owning your own business, because most new businesses fail in their first year." Most research projects do show higher costs and lower gains for rotational grazing, and most new businesses do fail in their first year, but this is due to management or the lack of it and not to inherent weaknesses in either business or grazing management.

*If you don't know what you are doing,
It doesn't matter a hell of a lot what you do.*

The knowledgeable and conscientious grazier will increase gain both per head and per acre while reducing both cost of production and total risk. The biblical axiom is true.

"The eye of the master fattens the ox."

The key is to understand that grazing management is a biological process with financial constraints involving many different factors and that the process must be constantly monitored and corrected to bring about the desired results. This sounds intimidating, but it becomes a straightforward procedure of checking to see if the actual growth rate corresponds to the growth rate used in the plan and if the stocking rate matches forage production. When problems are found, and they will be, the plan is changed to bring about the desired results. For example, if the plan calls for leaving a five-inch stubble after each early graze period and you find that the stock is consistently grazing down to three inches in the allotted time, you are likely overstocked and need to either reduce stock numbers or increase the area being grazed. If you find that the plants have not recovered in the time allowed between graze periods, you have misjudged the growth rate and need to allow more recovery time. The growth rate of forage is determined by the quality of growing conditions (temperature, moisture, sunlight, soil health, fertility), the rate of maturity of the forages (slow maturing plants need more time), and the amount of leaf left on a plant after grazing. Because so many variables are in play, any plan that is based on the calendar and not constantly monitored and corrected is almost certain to produce poor results. This is why research projects routinely show poorer results than are found by knowledgeable producers; good results require constant observation and fine-tuning of the grazing plan.

The value of forage to grazing animals is determined by the physiological age of the forage. Very young forage produces poor results, because, although it is highly digestible and high in protein, it lacks energy and enough fiber for healthy rumination. Old forage has more energy but is low in digestibility and protein. Forage that has had sufficient quality growing time to completely recover from the last grazing but has not yet begun to channel much energy into reproductive tissue has a good balance of protein, energy, digestibility, and fiber. Forage in this state also is normally tall and dense enough to allow animals to take in the required amount of forage with a minimum expenditure of energy as grazing work.

Planned grazing always involves locating animals in time and space, but it also entails matching the types and numbers of animals grazed with the available forage at the proper stage of growth and with the prevailing conditions of marketing, climate, finances, labor, and management. Having in place and monitoring a well-planned grazing program reduces risk and allows management to prevent crises rather than responding after they occur. A critical factor, too often overlooked, is stockmanship; Bud Williams type not PRCA, and what used to be called animal husbandry. If you want your animals to make you a profit, job one1 should be to ensure that they are as happy and content as possible.

The grazing plan might well be termed the operation plan, as, to be effective, it must take into account all aspects of the operation. Some in academia have complained that rotational grazing has not been shown by research to be superior to continuous grazing; I would suggest that there are serious problems with this reasoning. The first comes in the difference between research and management; the researcher lays out a protocol and follows it to see what occurs. The results of this approach are determined by how well the protocol fits the situation and by how the situation changes over time. It is impossible to lay out a rigid protocol that accounts for all possible changes in all of the factors that make up the situation. The results from a management program that does not allow for changes in the program to reflect variations in the factors that make up growing conditions will be at best substandard and are likely to be disastrous. To keep on course toward the goals, the good manager realizes that no plan is perfect and is constantly monitoring for faults in the plan and changing it when faults are found.

The manager starts with a set of desired results and manipulates the management process to achieve the desired results.

A second potential source of conflict is in defining what constitutes success; most research projects are short term, three years or less, and extrapolating changes observed in the composition of forage mixtures or forage production in the short term to the long term is dicey at best, as year effects play a large role in short-term changes in grassland. Animal production is easier to measure, but production is only one of the variables requiring attention from the manager.

Stability of production over time is at least as important to long-term profitability as is the amount of production.

Production taken at the expense of degrading the soil-plant-animal complex amounts to mining and can seldom be justified; an exception might to intentionally over-utilize forage in an area to weaken it and allow other forage species to be inter-seeded. If such a practice is used, plan to offset the damage with additional rest or whatever other remedy is appropriate. There are no hard and fast "thou shall not's" but it is critical that we understand the consequences of our management practices.

There are no "thou shall not's" in grazing, but understand the results of your management, and be sure these results are what you want.

Especially hard to evaluate are practices such as burning or weed spraying where the immediate results appear to be good; monitor the effects of such practices according to their effects on the ecological processes. The good manager will constantly monitor the results of his management on all aspects of the "whole" under his control; excellent results in one area may be more than offset by poor results in other areas. High animal productivity that comes at the cost of degrading any of the ecological processes will seldom be a good trade-off. A program in which production comes at high financial cost is always at risk and vulnerable to disaster caused by changes in climate, market prices, or costs of inputs. Good financial results that come at the expense of overwork to the point that health and family life are damaged are poor bargains.

To be valid, a decision must be financially, ecologically, and sociologically sound.

A major benefit of planned grazing is the ability to plan away the need for inputs and thus increase the margin between cost of production and sale proceeds. Stocking rate and income can be increased with high stock density grazing as both overgrazing and underutilization of forage plants are decreased. When forage is presented to animals at the proper state of growth through planned high stock density grazing, animal performance will be good with little or no need for supplementation; young, growing animals need young forage with a high protein content, while lactating or fattening animals require older forage with its higher energy content. These requirements can be met through grazing management by controlling the physiological age of the forage presented to the various classes of animals. By matching nutritional needs to forage quality, the differences in nutritional requirements of various classes of animals can be used as a tool to gain more value from low-quality forage and to increase the quality of a forage sward by removing older material. Frosted, warm-season forage grown under a high residual planned grazing program will be more uniform in physiological age, have more leaf and less senescent material, and be more valuable during the dormant season than the residual forage left after season-long, continuous grazing. If this forage is rationed out in one-day or less breaks to dry females, it can meet all or most of their nutritional needs and thus reduce the need for hay or other energy feeds.

Forage grazed or trampled down adds value to both the livestock and the soil and sets the stage for high-quality regrowth. In areas where pastures are burned in the spring to bring about stocker-quality forage, grazing off the old grass with dry cows at high stock density can increase income and build soil health, providing most of the benefits of burning with none of the bad effects. Occasional burning can be beneficial in special situations, but routine burning causes the loss of soil litter cover with the accompanying reduced water infiltration, increased water evaporation, reduced soil life, and increased weed encroachment. Burning also causes a loss of mineral nutrients that go off in smoke, blow or wash away, and are lost in the increased soil erosion brought about by having bare ground. Burning produces early high-quality forage but at the cost of an average 20 percent reduction in total forage produced in the year of the burn, and at the cost of species shift toward fire-tolerant plants such as little bluestem. Grazing the old material rather than burning it turns a negative into an asset. To be most effective, the forage should have been under high stock density grazing during the growing season, and graze periods on the dormant forage must be kept short; the animals should take all of the forage that is to be removed from the sward in one graze period. It is critical that enough forage be put on offer that the animals have some selection and are not forced to consume forage that has no value. The forage does not necessarily have to be consumed to bring about the benefits to the soil and to the forage; forage that is trampled into the ground will provide the same benefits to the soil-plant-animal complex as the forage that goes through the animals. Time spent moving temporary fence can be a very well paid job if the advance planning is done.

Planning is the most important work done by a manager.

Methods for Improving Animal Performance Through Planned Grazing

- Control the physiological age (maturity) of the forage presented to animals

The value of forage to a grazing animal relates more to the age of the forage than to the species of the forage. Forage that is too old has low digestibility and protein levels, while forage that is too young lacks energy and the fiber needed for proper rumen function. A weed at the proper state of maturity can be much higher-quality forage than alfalfa that is overly mature or too young.

- Match physiological age and, thus, the nutritional content of forage to the needs of animals

Understand that the nutritional demands of animals differ according to their physiological state, and use these differences to obtain maximum value from the forage produced. Weaning the calf of a heavy milking cow can reduce her need for energy by 35–40 percent, while increasing the body condition score of a dry cow from a 3 to a 5 will require about 35 percent more energy than she needs for maintenance alone. Plan to be able to present forage to the animals at a maturity state that best matches their nutritional needs; young, growing animals need the higher protein content found in immature forage, while lactating and fattening animals need the higher energy content of more mature forage. Ruminants have the ability to overconsume when forage conditions are good and to convert the extra energy into fat on their backs; this fat reserve, unless it has been genetically removed to fit the "demands of the trade," can allow animals to come through the lean times with very little help from man. It is wasteful and does no favors to the animals to maintain a high body condition year-round; ruminants are physiologically adapted to go through a seasonal weight fluctuation in sync with the seasonal changes in abundance of forage. The good manager will know when forage conditions are normally good and when they are poor and will design a breeding program and stocking inventory that match forage demand with supply.

- Present clean, even-aged forage at the proper state of maturity

The use of high stock density can provide this uniformity of age, while short graze periods ensure that the forage is clean and similar in nutritional content from day to day. As the forage sward becomes more uniform in age, animals will begin to utilize and benefit from a wider range of plants; many forbs that are seldom grazed in continuous graze situations are valuable sources of minerals and nutritionally valuable compounds not found in grasses.

- Present a dense sward of forage

A dense forage sward that is uniformly at the best balance of quality and quantity allows the grazing animal to use less energy harvesting forage, leaving more energy for growth, reproduction, and fattening. Grazing is hard work; the more uniform and dense the forage, the better the return to the grazing animal for the expenditure of energy as grazing and rumination. When using high stock density grazing, we take away some measure of the animals' ability to select their diet, so it is critical that the recovery periods be of proper length to suit the prevailing growing conditions and thus to present to the animals forage at the correct state of maturity. Animal performance is highly dependent upon forage dry matter intake; intake is increased by presenting animals with forages that are clean, dense, diverse, and of the correct nutritional balance.

- Present a forage sward of diverse species

No one species of forage is capable of providing the total nutritional needs of grazing animals. Forage species differ in their capabilities to take up various minerals with forbs and shrubs often having higher concentrations of minerals than grasses grown on the same soils; we have been badly oversold on the desirability of having "clean stands" of grass. The more forage species present in a pasture mixture, the higher the availability of nutrients at any point in time; in addition, the differing growth habits of plants in a diverse sward offer green forage over a greater percentage of the year. The more nutrition the animals get from grazing, the less need for supplements.

- Provide quality forage for as much of the year as is feasible

The opportunity to have green forage varies widely with climate, but even in areas with very short growing seasons, it is possible to improve the quality of the forage utilized by rationing. Grazing animals are just like humans; they always eat the best first. In a continuous grazing situation, the quality of the forage eaten is very high early in the growing season and becomes increasingly poorer as time passes. This is true even in areas with long growing seasons, as the forage on offer becomes a combination of overgrazed plants and senescent plants. By controlling the amount of area being grazed at any one time, forage grown during a short growing season can be rationed out, with each ration containing some leaf and some stem and the total making up an adequate diet. With a longer growing season, multiple passes can be made in each area timed to give the best value of quantity and quality.

- Match forage quality with nutritional needs of livestock and use livestock to promote forage quality

Grazing multiple species on the same area can greatly increase forage production. Sheep prefer forbs (weeds), goats prefer browse (brush), and cattle prefer grass. Weeds eaten by sheep not only do not compete with grass, but they become gain-producing forage and contribute to both energy flow and mineral cycling. Dry, mature cows, or other classes of stock with lower nutritional needs, can be used to clean up forage left by animals with greater needs. The forage available on the next grazing period will then be more uniform and of higher quality.

- Reduce stress on livestock with good stockmanship

Rule 1 is to make the animals as happy and content (stress free) as possible. Observe animal behavior, and learn to outthink them rather than trying to outmuscle them. Stockmanship boils down to understanding what really "must be done" and getting the animals to do it with the least amount of stress.

- Match animal production schedule with forage production

Forage should supply nearly all of the nutrients needed by grazing animals. The stages of animal production that require the highest plane of nutrition should be scheduled to occur at the times when nutrient production by forages is the greatest.

To reiterate, planned grazing differs from rotational grazing or any of the various grazing schemes in that planned grazing is results driven. The manager decides which things are to be accomplished and in what time frame and manipulates the available resources to make these things happen. There are always options; look realistically at the situation, and design a program that fits the resources and limitations present.

The plan may be useless, but the planning is invaluable.
                                    General Dwight Eisenhower

---

Grazing Plan Guidelines

The grazing plan is a combination calendar and road map; it will show which animals are in what areas at what time periods.

To make these decisions, we need to know:
- an approximation of the amount of forage in each paddock
- time frame of forage growth
- length of recovery periods needed by the forage being grazed
- an estimate of forage demand (stocking rate and stocking mix) for each day of the grazing season

Start with the stocking rate in present use or the stocking rate recommended by the NRCS for similar areas and create the stock flow chart as described earlier. As paddocks will differ as to the amount of forage grown, paddocks need to be ranked according to the percentage each contributes to the total of forage produced; this ranking can be done by clipping and weighing forage from equal sized areas in each paddock but for planning purposes (that will be carefully monitored and corrected) approximations will suffice (paddock 10 is the strongest and rates 100 percent or 1.0, paddock 4 grows 60 percent as much forage as paddock 10 and so has a relative rating of .6, paddock 3 grows 80 percent as much as paddock 10 and is rated .8, so on and so on). If the grazing season is the entire year, the animal units present will be distributed through the paddocks for each day of the year according to forage availability. In its simplest form, the grazing chart is a calendar with paddocks listed down one side allowing the manager to note that herd X was in paddock Y on day Z. It should also have provisions to note when events such as calving, weaning, shipping and, breeding will take place. Any paddock that will not be available to graze on a given date should be marked on the chart as being unavailable at the appropriate dates. If any paddocks require different recovery periods, these should be noted. HMI has excellent grazing charts available and clear instructions as to their use in *Holistic Management Handbook* by Butterfield, Bingham, and Savory. The handbook and charts along with the text book *Holistic Management* are available at www.holisticmanagement.org.

- Determine the number of animal unit days of grazing available from each paddock for the entire year

    The goal is to determine where animals will be at which times and for how long. The simplest way to do this is to rank paddocks from 100 percent for the most productive to whatever percentage rating fits each paddock according to the amount of forage produced per acre. Divide the relative production rating of each paddock by the average relative production rating (total of paddock relative ratings divided by number of paddocks) and multiply this figure by the average AUDs provided by each acre for the year (total AUDs required divided by total acreage). This gives the AUDs supplied per acre/year for the

paddock in question. Multiply AUDs per acre time's acres of forage in paddock to get AUDs/year for the paddock. Divide this figure by the number of AUs present to get the number of days this paddock will be grazed in a year.
Example:
200 acres divided into ten paddocks and carrying fifty Animal Units for the year
365 days per year X 50 AUs = 18,250 AU days required
18,250 AUDs / 200 acres = 91.25 AUDs per acre per year average
For this example assume that the average of all of the paddock relative production ratings is .81
Paddock 6 is thirty acres with a relative rating of .75 (it is 75 percent as productive as the best paddock)
(.75/.81= .93), 91.25 average AUD/acre X .93 = 84.5 AUD/acre for Paddock 6.
84.5 AUDs X 30 acres = 2535 AUDs per year in Paddock 6.
2535 AUDs divided by 50 AUs = 51 days paddock 6 will be grazed in one year.

- Decide how many of each class of livestock are to be run; include any wild grazers

  This information comes from the stock flow plan.

- Decide how many herds are to be run and the number of animals in each herd

Keep the number of herds as low as possible to increase stock density and to reduce costs.

- Decide whether or not a leader-follower program is to be used, in which one herd closely follows another through the paddocks

    A leader-follower program can be very useful, but care must be taken that the follower herd is close in time to the leader herd; the recovery period should be figured as the time from when the follower herd leaves the paddock until the leader herd arrives back in the paddock.

- Decide on the range of lengths of recovery periods to be used

    Many operations can benefit from using recovery periods of three lengths: the shortest length should be used when growing conditions are best and forage is growing the fastest, medium length when forage is growing slowly, and longest when dormant forage is being rationed out in a one-time over utilization. The absolute lengths of the various recovery periods are estimated using the criteria spelled out in "4. Recovery periods:" under "Factors Controlled in Grazing Management" in this chapter; these lengths are then refined by observation. The shortest recovery period can be morphed into the medium length period, but the longest recovery period is really just a rationing period. The length of time needed for new grass to become available, the number of animal units that are to utilize the dormant grass, the amount of grass available, the animal performance required, and the length of the graze periods to be used will determine how long and how heavily each section is grazed. Using temporary fence and short graze periods can be valuable in maintaining animal performance and reducing the cost of dormant season feed. It is valuable to animal performance to have old grass remaining and mixed into the diet (older forage can provide needed energy) until the new grass has time to achieve phase 2 (balance between protein and energy) status.

The need to consistently feed large amounts of hay or other energy feeds during the dormant season is a signal that beneficial changes in management are possible.

- Decide the number of days that each paddock must be grazed during each cycle to provide the required recovery periods

    Divide the number of days in the shortest recovery period by number of paddocks available to obtain the average number of days grazed per paddock. Example: 45 days recovery period divided by the 29 paddocks being rested in any one day equals 1.55 days average per paddock. If there are significant differences in the amounts of forage in various paddocks, the days per paddock must be adjusted to reflect these differences. Example: paddocks 1–24 have about the same amount of forage, but paddocks 25–28 have 75 percent less than average forage, and paddocks 29–30 have twice the average amount of forage. Paddocks 1–24 would be planned to be grazed 1.5 days per paddock for a total of 36 days, paddocks 25–28 would be grazed 1 day per paddock for a total of 4 days, and paddocks 29–30 would be grazed 3 days per paddock for a total of 6 days. This would yield 46 days minus the 1.5 days spent in the first paddock for a total of 44.5 days recovery, which is close enough for government work. In reality, graze periods could be made on full-day amounts, with paddock 1 grazed one day, paddock 2 grazed two days, paddock 3 grazed one day, and so on. If this is done, care should be taken that the paddocks shorted on recovery during the first cycle are given extra rest on the next cycle. It is critical that forage use be closely monitored to find the mistakes that are sure to be made in estimating the amount of forage in the various paddocks. Repeat the process for the other length of recovery periods used, and plot out the distribution of animals in time and space for the growing season. The length of graze periods used will greatly influence animal performance and soil and forage improvement; if no one is around except on weekends, it is worthwhile to try to find someone to come in once a day to move animals and check water.

Done correctly, short grazing periods and the resulting high stock density are extremely valuable for promoting plant and soil health and for improving animal performance; this same high stock density, however, takes away much of the animal's ability to select its diet, so it is critical that sufficient quantity and quality of forage be provided. Monitor the amount of forage in the paddocks both before and after grazing, but also monitor animal performance; if animals are unhappy or frantic to move, something is wrong, and the problem must be found and corrected. Moving to a fresh break of pasture is great fun, and animals will learn to lie to promote an early move; it is the manager's job to separate fact from fiction.

- Look carefully at the possibilities of using planned grazing to prevent problems before they occur and to alleviate preexisting problems

Areas that normally accumulate deep snow can be grazed before snowfall, while areas that normally have less or no snow are saved to be used during and after storms. Even where winters are not severe, having an area available with a windbreak and forage can be a real stress reliever when a "blue norther" comes roaring over the horizon; it helps if the animals can either be gotten to shelter before the storm hits or be on the north side of the hidey hole.

Areas with weak water should be scheduled to be grazed when water is most plentiful; conversely, wet areas should be scheduled for use when it gets dry. The same reasoning holds for any event or condition that can be predicted; it is probably smart to not have the cattle (and especially the sheep and goats) mixed in with the deer and elk when hunting season opens, and it is usually better to hold off grazing the loco weed pasture until loco is not the only thing green. Areas where there is considerable danger of fires starting, such as along a railroad or highway, should be grazed as soon as wildfire becomes a danger. As I write this, much of Texas is burning; we can't cure drought and high winds, but with thought, we can reduce their effects.

Financial Plan

Many farmers and ranchers suffer from a severe allergic reaction to pencils and paper; I have known men who could sit in a sale ring for hours buying cattle for six different accounts and could tell you at any point how many animals went to each account, what they weighed, and what they cost—but they could not balance their own checkbook.

> *To be sustainable, an operation must be profitable; no operation can be truly profitable without being sustainable.*

I have covered most of my thoughts on finances in chapter 5 but will elaborate some here. A financial plan is a reckoning between what we expect to come in and what we expect to go out; when planning out the financial year, rule 1 should be—be realistic with the figures. This is particularly important when operating on borrowed money; be conservative estimating income and liberal estimating expenses. If there is going to be a big difference between what you plan and what actually happens, you want it written in black ink, not red ink. The actual plan does not have to be complicated, but it should be capable of providing close estimates of income and expense by enterprise (see chapter 5) for each month of the year, and it should have a way to compare what was planned with what actually happened.

Example:

	January		January	
	Expense Expected	Expense Actual	Income Expected	Income Actual
Profit				
Equity increase (debt reduction)			1500	1500
Cash (draw + reserve)			1800	1750
Overhead expenses				
Mortgage reduction	1,000	1,000		
Op loan reduction	500	500		
Cash reserve	300	250		
Management draw	1,500	1,500		
Pasture rent	8,000	8,000	0	0
Insurance	130	130	0	0
Taxes	6,000	6,000	0	0
Labor hired	1,200	1,200	0	0
Truck expense	375	360	0	0
Repair & maintenance	300	218	0	0
Total overhead/month	19,305	19,158	0	0
Capital expense				
Fencing	0	165	0	0
Water development	200	0	0	0
Other	0	0	0	0
Total Capital/month	200	165	0	0
Enterprise expense & income				
Cow – calf	725	675	0	675
Replacements	200	175	0	0
Home raised stockers	150	120	0	0
Purchased stockers	0	0	0	0
Hay	0	0	0	0
Total heading/month	1,075	970	1,050	1,625
Actual	1,075	970	0	675
Total cash	20,580	20,293	0	675
Cash Difference		-287		675
Total In house			{2,300}	{2,250}
In house difference				{-50}

Each heading would have an accompanying work sheet giving the specifics of both expenses and income for each month. The expenses listed under **Profit, Overhead,** and **Capital** must be paid out of the margins between income and expense generated by the various enterprises. Computing this margin for each enterprise is known as a gross margin analysis, and it will show what each enterprise contributes to or subtracts from overhead expense. It is not unusual to see operations change from losing money to becoming profitable after identifying and eliminating one or more enterprises that are losing money and dragging down profitability for the whole operation. To be useful, the enterprises must be described in detail; cattle, crops, and hunting won't cut it. Within such general categories, wheat may be a winner while corn consistently loses money, and feed to fatten sale bulls should be charged out as sale bull feed rather than simply as cattle feed. The point is to specifically identify the enterprises that make money so that more resources can be devoted to these enterprises and none to those that lose money.

It is not hard to make money, but it is darn hard to keep from spending it.

Financial records, like all records, are useful only if they are used; the purpose of keeping records is to provide the tools needed to make good decisions. The data listed on the above form can be used to make gross margin analyses on the various enterprises and thus give the information needed to decide what resources should be devoted to which enterprises in the future.

I wrote the following in 1994, but it is still true today.

Rules of Hard-time Economics

1. Expenses always rise to meet the level of disposable income. Now is the time to reassess the difference between "we need" and "we want."
2. Maximum production is seldom (if ever) as profitable as some lower level of production due to the laws of diminishing returns and marginal reaction.
3. Ranching is not the business of producing livestock but rather the business of converting solar energy into biological energy and thus into money.
4. The land (defined as the total soil-plant-animal complex) is the source of all new wealth. The ability of the land to produce wealth (except for mining operations) is in direct proportion to its biological capital. Biological capital is biodiversity plus the long-term effects of biodiversity. Biological capital consists of healthy populations of healthy individuals of many species of plants and animals fully utilizing the available resources—and in the process, improving the resources.
5. Get your program in tune with nature, and take advantage of all available freebies. Design your operation so that the period of greatest nutrient demand coincides with the period of most nutrient supply.
6. If you are producing a commodity, to consistently make a profit, you must be a low-cost producer.
7. All operations consist of multiple enterprises (any endeavor that produces or consumes income). Analyzing all enterprises and discarding the unprofitable ones can mean the difference between success and failure.
8. Most people do not succeed, so doing what is accepted as the norm is a recipe for failure. Most people also do not

wish to see others succeed and are quick to ridicule any departure from standard behavior.

9. Agriculture is the most overcapitalized industry in the United States. Most ranches have far too much money tied up in machinery and improvements.

10. Much of the money spent on pest and disease control is wasted, and often the cure is worse than the malady. Most problem organisms are symptoms of poor management. Any practice that destroys life should be carefully monitored to determine whether its effects are the ones desired.

11. Stability of production is at least as important to consistent profitability as is the level of production. Stability is a function of biodiversity, and the operations that survive the "drought of the century" and the "market wreck" will be the ones that store up biological capital during good times.

12. All agriculture is inherently risky, as it is subject to factors such as weather, markets, and politics, over which the producer has little influence. When profit margins are thin is a poor time to increase inputs in the hope of increasing production. Increased inputs at this stage are valid only if they decrease the cost per unit of production or reduce the likelihood of production failure.

13. All life forms have a set of environmental circumstances to which they are adapted. Do not attempt to move plants or animals out of their area of adaptation—or even worse, attempt to modify an area to suit non-adapted plants or animals.

14. Planning is the most important work that a manager can do. Planning production is fun, but planning profitability is more rewarding. Pay yourself first; plan profit into

your program just as you plan to pay the mortgage. Most crises result from a lack of planning. If you are a good problem solver, you are a poor manager. Good managers plan to prevent problems and don't get enough practice at problem solving to become good at it.

15. Realize that Con-Agra doesn't care whether or not you make a profit. Produce the cattle that work under your conditions, not cattle that "fit the box" or "meet the demands of the trade."

16. Understand that the feedlot industry has nothing to do with the cattle business. It is solely concerned with selling cheap grain and renting lot space. The packers and feeders don't want a ruminant capable of thriving on grass. They want a pig in a cow suit that can eat thirty pounds of corn, gain five pounds a day, and dress 70 percent.

Someday, I may attempt an entire book on planned grazing; here, I have attempted to describe some of the benefits and the rationale of planned high stock density grazing without going too deeply into the mechanics. *Holistic Management Handbook*, by Butterfield, Bingham, and Savory, is an excellent source for this information and much more. I strongly recommend the use of the Holistic Planned Grazing Chart; hopefully, in the near future, we will have a software program available to speed up the charting process.

CHAPTER 8
Substitute Management for Money

In a search for efficiency and labor saving, all of agriculture spends far more on inputs than can be justified by the potential for profit. The concept of best management practices is particularly flawed in that it assumes that if each aspect of the operation (weed control, soil fertility, nutritional supplementation, vaccination program, and so on) receives all the inputs it needs, the whole must function well. This concept is a recipe for financial disaster, as it judges expenditures based on their effects on one particular aspect of the operation and on whether these expenses will cash flow. To be valid, an expense must cash flow; however, even if it does cash flow, the expense may or may not be the best use of those dollars at this time.

A much better set of criteria for judging an expense would be:
- The benefit of the action to the whole operation
- The benefit of using the money for other purposes
- The cost-to-benefit ratio of the action
- The amount of money put at risk

- The degree of that risk
- Whether the expense will have to be repeated

All agriculture is inherently risky due to factors such as weather, economic changes, and politics, over which producers have little control. The more dependent a program is on outside inputs, the more vulnerable it is to the radical changes in weather, markets, and costs of inputs that are certain to occur. Whenever spending money is contemplated, the first question should be, "If this purchase is made, will it solve a problem in a permanent and financially feasible manner, or will it only treat symptoms of more basic problems?" If a purchase or action will have to be repeated in the near future, it is probably a poor use of money and possibly signals an opportunity to make a beneficial change in management.

Adding inputs makes sense only if it will increase profitable *production or increase the stability of production.*

Many of the so-called pasture improvement practices fail all parts of this question. They address the symptoms of weed and brush encroachment and loss of forage species rather than the real problems of degraded ecological processes and poor grazing management. Chemical or mechanical brush and weed control measures are good examples. Benefits of these practices last about as long as a politician's promise, and most of them make the situation worse in the long term by simplifying the soil-plant-animal complex (the grazingland). In most cases, both the ecological and financial costs are higher than is justified by the short-term benefits. When controlling costs, the question should not be "How do I lower the cost of this practice?" but rather "How do I change the program to get rid of the need for this practice?" Listing all expenses according to the enterprises and practices that use the money is an excellent way to start this process. Sometimes the answer is to drop entire enterprises; if you have to feed four tons of hay per cow per winter on the High Lonesome, maybe you shouldn't have cows on that property. Allen Savory is correct when he says that it is possible to be doing everything you do very well while doing exactly the wrong things.

A great deal of the time, effort, and money that goes into farming and ranching operations is spent trying to clean up messes that we have created with poor management practices. I would explain that statement by examining our management practices in the following categories.

- Weed and brush control
- Insects and other animal pests
- Forage production
- Supplemental feed
- Animal health
- Animal productivity

  - Weed and brush control
    Control of weeds and brush on pastureland is a multibillion-dollar-per-year business in the United States. Techniques range from hand chopping to biological warfare using fungi and insects. One of my least favorite chores as a kid was to walk the creek

with a tow sack and pull up, roots and all, cocklebur plants so the burrs wouldn't get in the wool and mohair of Dad's sheep and goats; it didn't matter whether or not the plants had burrs, I pulled them up and hauled them to the bare ground of the caliche pit and burned them.

Wherever a road or cow trail crossed the creek, there was always a patch of cockleburs, and it was my job to root them out before they swarmed out of these enclaves and took over the entire pasture. Dad didn't like cockleburs; he would probably let a sheep-killing coyote go before he would neglect a cocklebur. It was not until much later that I finally realized the only place cockleburs grew was on disturbed or bare ground; if I could figure a way to heal bare ground, I could win the great cocklebur war. This is the key to all pest control: you don't have to kill the pest; you just need to change the conditions that are favoring the pest.

If you don't like what is growing in your pasture, change the growing conditions.

Hundreds of thousands of acres of brush land are sprayed, mowed, root plowed, grubbed, bulldozed, burned, plowed, or poisoned every year, but the brush and weeds continue to get worse. Huge areas that were healthy grassland fifty years ago are now brush thickets and weed patches. Like most ranchers, I have been to dozens of extension and chemical company meetings where the theme was how to kill weeds and brush, but I have never heard these people give a satisfactory explanation as to why the weeds and brush are expanding their range. Weeds and brush gain a stranglehold on pasture, because our management gives them the advantage over the forages we are trying to grow; it really is that simple.

Weeds and brush increase, because our management favors them over the forages we are trying to grow.

One of the buzz words in range management today is invasive. In a recent issue of *Rangelands* the entire magazine is devoted to articles about invasive species. Reams of paper have been used both in the popular press and in scientific journals detailing the damage done by invasive pest plants, everything from natives like mesquite and eastern red cedar to introduced plants such as leafy spurge and diffuse knapweed. Large amounts of time, effort, and money are spent in efforts to control these invasive species, but long-term success is rare. It is undeniable that we have lots of pest plants in our pastures, that they are becoming more common, or that they reduce production and profitability. I do quarrel, however, with the concept of a plant being invasive.

There is no such thing as an invasive plant; there are only plants that are adapted to a certain set of environmental conditions. If a kind of plant that you don't want is increasing in your pasture, it is not because that plant has some special characteristics that give it invasive powers; it is because the conditions in that pasture favor the pest plant over the more desirable plants. (I can almost hear the screams of outrage at this, and this has not even been published.) I realize that the black magic hoodoo weed was introduced from the back side of Lower Slobovia and has none of its natural predators to hold it in check, and I know that the pink and purple cow eater is toxic and nothing will eat it—but this does not change the fact that these plants are able to increase in our pastures only because our pastures are not healthy. We have plenty of native plants that have their full complement of natural pests and are still increasing fast enough to be termed invasive. Why, after being part of the local environment for eons, have these plants suddenly turned invasive? It is not due to some change in the plant; it is due to changes in the local grassland environment. These "invasive" plants lose their invasive ability when confronted with healthy grassland. We don't like to admit it, but we create, by our management, most of our own problems, whether they are weed and brush encroachment, lack of profitability, or everything in between.

Weeds are not caused by a 2-4-D deficiency.

Perhaps the most common management fault promoting weed and brush intrusion is continuous grazing, which is always detrimental but is a guaranteed range health and profit killer when combined with too high of a stocking rate. We have seen some very good work lately that details how expensive hay is for beef cattle, but another expense of the "stock heavy and feed hay for six months" program is often overlooked. Forage plants that are defoliated too often and too severely lose vigor and the ability to compete with weedy plants that are not grazed as heavily. When the number of species of plants on grassland is reduced by abusive grazing or by chemistry, unfilled ecological niches are created, and the area becomes less complex and more unstable. Land that is grazed so heavily (or hayed or burned so often) that the soil surface is not covered at all times with vegetation (either alive or dead), it is subject to wide temperature extremes and loses soil life. Such soil is less able to take in or hold water and becomes a hostile environment for high-quality forage plant seedlings. Nature hates bare ground and the waste of sunlight, moisture, and mineral nutrients that occurs when ground is bare, so she sends in plants (weeds) that can survive in the harsh conditions.

We create unhealthy grasslands with abusive grazing and then compound the problem with "cures" such as chemicals, tillage, and burning, creating more simplification and more bare ground, which makes the problem worse.

Weeds, both common, local types and the imported super weeds, are nothing more than nature's scabs attempting to heal what is damaged.

Weeds can establish and grow on degraded areas where our favored plants cannot, and they can survive abusive grazing. If we don't want weeds, the logical approach is not to kill the weeds but rather to change the local growing conditions to favor the type of vegetation that we want to grow.

There is another aspect to promoting desirable plants; many years ago when I took range management courses, we were taught that plants were either palatable or unpalatable and were increasers or decreasers under continuous grazing. That is a pretty easy concept to understand; cows eat those plants that they like and leave those plants that they don't like, and before long, there is a pasture full of plants that they don't like. The situation is actually a little more complex than that; palatability of plants to grazing animals involves much more than the taste of the plant. As explained earlier, palatability is influenced by several factors; physiological age of the plant is important as is the percentage of the diet made up of the plant and the availability of plants with offsetting antiquality factors.

As an example, little bluestem was termed a decreaser and very palatable, yet in continuously grazed pastures, especially in hard-to-graze areas, there are always "wolf" plants of little bluestem that livestock have refused to eat. Even though it is a valuable forage plant (one of the "big four," in fact, of the tallgrass varieties), little bluestem matures early; any plant that is not grazed during the spring flush when forage is most abundant will get tough and will be refused from then on by any but a starving animal. These "wolf" plants take up space, sunlight, water, and nutrients but contribute little to the health and productivity of the pasture sward. As a species, little bluestem is palatable, but these lignified individual plants are not. Timing and degree of defoliation are very much a part of "growing conditions" and are among the easiest of conditions to control.

The degree of "palatability" of a plant is also very much influenced by the concentration of that plant in the forage on offer and the amount of diversity in the pasture sward where it is growing. (How many more biscuits will you eat if you have cream gravy or butter and wild plum jelly than if you have nothing to eat with them?) Go to www.behave.net to learn more about what makes one plant more palatable than another in various situations.

What is needed is not stronger weed sprays or more effective ways to kill brush but rather management that promotes the total health of the soil-plan-animal complex.

- Insects and other animal pests

Controlling insects and other animal pests is another area where, through our management, we have been very successful in making the problems worse. In 1962, armed with a brand-new degree and all the latest technology, I took over management of a ranch in the Red River Valley of southeast Oklahoma and set out to remedy for all time the problems of horn flies, internal parasites, heel flies, pecan insect pests, and goat-killing coyotes; I got my tail kicked on all fronts. I set up a program in which we would spray the cattle for horn flies every 26–28 days from sometime in April until we had a killing frost; in addition, we wormed the cowherd twice a year, the calves and yearlings at least as often, and the goats three to five times a year. We sprayed our pecan trees at least twice and usually three times; when we weren't spraying cattle or pecan trees, we sprayed weeds. I set out a trapline with traps, snares, and cyanide guns, and I shot coyotes both in the daylight and at night and even roped one. I killed stomach worms, the occasional goat or calf, horn flies, heel flies, pecan casebearers, aphids and coyotes, along with most of the beneficial insects, all of the forage legumes, and any chance of making a profit. I also damaged my own health and that of everyone around me; I shudder when I think about some of the stuff I brought in on my clothes to contaminate my wife and little girls. I was in the classic mode of managing against what I didn't want, and I stayed in this mind-set for years.

I would like to say that, realizing that what I was doing was not working, I analyzed the situation and devised a better plan of management—but that would be a lie. The cattle market crashed, and it was painfully obvious that I had to change my ways or we would go bankrupt; necessity really is the mother of invention, and I started trying to find ways to keep from buying inputs. Cold turkey, I stopped using nitrogen fertilizer, weed sprays, and insecticides of all kinds; we used no fly tags, wormers, back pours, or pecan spray, and we worried about the wreck that we were sure was coming. We had been using rotational grazing for years, but with only five or six paddocks per herd and long graze periods. I began to build more subdivisions as a way to increase stocking rates, and we saw good things happen with pest control when we got our stock density up and quit using poisons. Earthworms began to appear where we had never had them before, and dung beetle numbers exploded to the point that when it was warm and moisture was present, manure would be buried or desiccated within a few days. Horn fly, heel fly, face fly, and horsefly numbers all declined, and beneficial insects such as sand wasps and spiders increased. With the ability to control the time animals were on an area came the ability to control the height of the forage left behind; if we left at least a three-inch stubble, we left the vast majority of stomach worm larvae behind, as most of them do not crawl higher up the grass than about two inches and so were not ingested by the cattle.. The combination of frequent moves to clean pasture and leaving a high leaf residual completely did away with the need to worm cattle; I did still worm anything that I brought in from outside, but I saw no need to worm our home-raised cattle.

I waved the white flag of surrender to the coyotes and shipped the goats back to Texas, but years later, when I brought sheep back into the program; I was able to do the same worm control with sheep. I turned the coyote problem over to four good guard dogs, and they did a much better job than I ever had. I would pick up an occasional wormy sheep that I would treat and then sell, but I did not treat the whole flock for five or six years. Sheep and goats are afflicted by internal parasites just as cattle are, but by different parasites; a sheep parasite picked up by a bovine will die as will a cattle parasite picked up by a sheep or goat. Thus, by grazing multiple species over the same ground in rotation, it is possible to drastically lower the parasite load for all species. Notice that I did not say that my animals had no internal parasites—just that they did not need worming. Animals, humans included, are supposed to have parasites; however, they are not supposed to have so many that life processes are threatened. If a treatment kills all or most of the parasites in an animal, the animal will stop producing the antibodies that protect them from parasites, and the next time the animals are exposed to parasites, they will pick up a heavy load.

The way to control internal parasites and most other parasites is with a multiple-method approach:

- Reduce the parasite load in the soil by encouraging soil life with high stock density
- Move the animals frequently to clean forage
- Use rest periods that are as long as is feasible, with no susceptible animals present
- Keep the animals grazing up off the ground
- Use multiple species in the stocking mix

- Build healthy immune systems with proper animal nutrition and low-stress handling
- Include high-tannin plants in the pasture sward, as tannins prevent the implantation of parasites in the gut wall
- Actively select for genetic resistance to internal parasites, lice, and horn flies
- Use chemicals only as a last resort; otherwise, you may win the battle but lose the war

  - Forage production

  During World War II, the United States built a lot of capacity to produce ammonium for use in explosives; when the war ended, this material was funneled into the fertilizer market, and nitrogen fertilizer became both abundant and cheap. When cheap nitrogen was lavished on Bermuda grass in the South and tall fescue in the North, forage was produced in unheard-of abundance. I remember a farm magazine sponsoring a contest to see who would be the first to produce a ton of beef on an acre of grass. I don't remember if anyone got the prize, but nitrogen fertilizer quickly became firmly established as an absolute necessity on tame pasture. This mindset continued even when prices for nitrogen rose as demand caught up to supply; the nitrogen frenzy continued even when it became evident that serious problems beyond the cost of nitrogen existed. High rates of nitrogen fertilizer burned soil organic matter up at a rapid rate, and soils lost life and tilth. Soils

became more acidic and began to show deficiencies in minerals that historically had been in sufficient supply. Forage production was concentrated in the most favorable growing months as soil water was depleted by rapidly growing forage; energy and mineral content of the forage fertilized with nitrogen dropped. It now took two or three times as much nitrogen to obtain the same plant response, and strange things began to happen to both plants and animals. People began to see nitrate poisoning of animals, an unusual amount of eye problems, soft tissue infections in ruminants, and plant diseases diagnosed as trace mineral deficiencies. The decay cycle slowed dramatically over large areas of both pasture and cropland as soil life was reduced; mineral nutrients were lost as plant material and manure decomposed chemically instead of biologically.

Highly available nitrogen fertilizer acts on the soil-plant-animal complex exactly like amphetamines act on the human body.

Among the first symptoms of a soil sick from loss of biological activity is a breakdown of the decay cycle and the appearance of trace mineral deficiencies. In a biologically active soil, microbes of all sorts are constantly at work decomposing dead organic material and breaking out minerals from this material as well as from the elemental forms found in rock. The products of these activities are sequestered in organic forms that slowly become available to higher plants. In addition, various microorganisms, such as mycorrhizal fungi and rhizobia bacteria, form synergistic relationships with higher plants; these relationships greatly benefit the plants by increasing the supplies of water and nutrients available to them. In a soil with high biological activity, plants also benefit from a reduction of organisms that cause disease and parasitism. Soil life in a healthy soil is diverse and complex, with many different types of microbes living in associations that are simultaneously mutually beneficial and self-limiting. In a situation with high biodiversity, due to competition, no one organism can produce population numbers high enough to reach pest status. As soil life is killed off, however, the system of checks and balances is lost, and plant health suffers. When plant health suffers, animals (including humans) that are dependent on those plants suffer as well.

The way to decrease plant disease and animal parasitism is not to sterilize the environment but rather to encourage healthy and diverse microbial life.

As experience was gained, a few lonely voices began to question the use of high rates of nitrogen, but most of us (including me) just spread more lime and potassium on our land, added vitamins and trace elements to our animals' diets, and bitched about the cost of all the stuff we were buying. Weeds loved the high nitrogen-degraded soils, so weed pressure increased markedly, and we were advised to use increased amounts of more expensive weed poisons. In Bermuda grass areas, quality forage was produced from late spring until mid to late summer, so with the cool-season legumes done in by nitrogen and weed spray, people began to feed more hay for longer periods of time; we could grow twelve-thousand pounds of forage per acre, but it all grew in ninety days and did not weather well. Farther north in the fescue belt, people began to talk about the "summer slump" when animal performance went to pot, because in the heavily nitrated and weed-sprayed pastures, there was nothing for them to eat in the summer except overly mature, endophyte-infected tall fescue. Even in the native grass areas, people tried to use nitrogen on native grass, and when that didn't work, some plowed up native grass to plant sudangrass or fescue. We got caught up in the "more is better" syndrome and forgot that the point should be to convert solar energy to biological energy to wealth as effectively as possible; the amount of forage grown is not as important as the total cost of the forage eaten by the animals. From a forage management standpoint, *the goal should be to maximize the number of days that animals need nothing but the forage that they harvest by grazing*.

- Supplemental feed

Supplemental feeding of grazing animals is not a big deal; it entails some salt and minerals and maybe some protein during the dormant season. The problem comes when supplemental becomes substitution; the economic advantage of ranchers is in turning something of low value (grass, and especially grass growing in the pasture) into something of high value (beef, lamb, and so on). Whenever we substitute grain or stored forage for forage harvested by the animal, we reduce this economic advantage; the need to substitute feed for grazing should be a red flag of warning of the largest size.

There are several reasons that people feed energy to grazing animals on pasture, but by far, the most common is overstocking; when you have too much stock for the amount of forage you grow, nothing is going to work right. An overstocked range will produce far less forage than a similar range that is stocked correctly; it will also produce far less profit and a lot more headaches. The second most common reason to need substitution feeds is poor grazing management; continuously grazed areas or areas grazed with long graze periods produce less forage and have lower grazing effectiveness. A third reason is poor energy demand management; in this case, because of either animal numbers by type or the physiological state of the animals being grazed, more quality forage is needed than is available; you have plenty of dry cow quality forage but not enough forage of stocker quality. The energy demand of a cow almost doubles when she gives birth and starts milking; if she calves in January and all she has to graze is tromped-over, dormant, warm-season grass, she is going to need a bunch of help.

Snow cover is a reason frequently given for substitution feeding, and at times it can be absolutely necessary; there are people who ranch in heavy snow country, however, who feed little or no hay. Cattle can learn to graze through pretty heavy snow if there is something under there to make it worthwhile; this is a learned behavior that they are not apt to acquire standing on a feed ground. I sat in on a blog discussion recently on this topic, and several ranchers in heavy snow areas agreed that their animals were healthier and stayed in better shape if they could keep them grazing even in deep snow; their observation was that when cattle received hay intended to supplement their pasture, they stopped grazing and lost condition. These ranchers were experienced, cold-country graziers, and they planned for snow; they grazed the areas where they knew drifts would form early, and they saved the slopes and areas that they knew snow cover would be lighter.

I have seen the same welfare cow phenomena even when there was no snow on the ground: cattle doing fine until somebody decides that "they need a little help" and brings out some hay; in my experience, trying to "stretch pasture" with a partial feed of hay seldom works. The exception would be four or five pounds of very high-quality hay fed as a protein supplement.

I would suggest that you ration out forage using temporary fence, break the ice crust with a tractor, and do anything else you can to keep animals grazing as long as there is quality forage available; when the forage is gone, put the animals on a full feed of hay in a sacrifice area. There is one more reason for needing substitution feed; many of our cattle are too big, give too much milk, have too little body fat, and are in general poorly suited to producing by harvesting their total diet by grazing.

If you want to be profitable, use animals that have the ability to produce efficiently on what your ranch has to offer.

- Animal health

Animal health is big business, and an amazing number of health aides are on the market; you can buy vaccines, antibiotics, bloat controllers, insecticides, vermicides, fungicides, and nutritional supplements to treat any conceivable aliment—or you can improve your management and buy a whole lot less. For many years, when it was time to wean, I separated the calves from the cows and locked the poor little devils away behind a stout board fence while the cows bawled, frantic to find their calves, and dashed back and forth outside. The calves would walk the fence and bawl until some of them lost their voices, and they would lose weight; they all got sick to some degree, and some got very sick. We gave vaccines to prevent sickness before we weaned, and we treated sick animals with antibiotics and probiotics (and would have used voodoo if we had known how)—and still we lost calves. I stopped weaning in this way when Steve McCune, a good Kansas rancher, talked me into weaning in the pasture using an electric wire; since starting to use this method, I have weaned thousands of calves with very little sickness, no medications, and no death loss. Calves don't get sick at weaning because of a lack of milk; they get sick because they are stressed out of their minds by fright, dust, change of diet, and anxiety. When we wean in the pasture on good, preplanned grass, the calves can touch noses with momma through the fence and fill their bellies with the same thing they have been eating. They can lie down and sleep right across the fence from momma; they would like to have a sip of milk, but they are not hungry, and they are not scared, because momma is right there, and she is not scared. The stress is gone, and instead of being a traumatic event in which calves get sick and lose weight, it becomes a process in which calves settle down and gain weight during the weaning process.

We cause most animal health problems with stress of one sort or another. One time I sold a widow lady twenty bred heifers that we gate cut out of a bunch that we were calving; she took them home and put them in a little trap where she could watch them—and watch them she did. Every time a heifer would lie down to strain, she or the neighbor (or both) would be right there in the heifers' space to "help"; shortly after that, she would call me to come pull the calf. On the 380 heifers we calved at home, we pulled maybe five or six calves; conversely, from the widow lady's twenty heifers, I, or someone else, if she couldn't find me, pulled nearly every calf. Regardless of good intentions, having a human in an animal's space makes the animal mentally uncomfortable, and if animals are uncomfortable, they are in stress.

Maybe second to psychological stress in causing animal health problems is poor hygiene; parasites and disease organisms build up in areas where their host animals are constantly present. At one time, I would ride heifers out of the herd when I thought they were close to calving, and I would put them in a calving trap where I could see them more easily; as the heifers calved, I would ride the 2–3-day-old pairs out and bring in new "piggies." I took one set of heifers and made three groups out of them, thus greatly reducing the number of paddocks available to each herd; I was forcing the heifers to have their calves on contaminated ground, and it showed in the number of calves sick with infected navels or scouring. I had a lot of calving difficulty, because I was still hung up on having heavy weaning weights and selecting for heavy cattle, and I was calving in the winter so the calves would be big enough to use the grass when it arrived. I don't think I could have done things any more wrong if I had set out to create a disaster. When I finally got my act together, I learned that having baby animals born on clean green grass to mothers that are in a good nutritional state, have the right genetics, and are physiologically ready to give birth is a pleasure to man and beast.

- Animal productivity

Farmers and ranchers are competitive people and take a lot of satisfaction from big crop yields and heavy weaning calves. One of the quickest ways I know to start a fist fight is to tell a group of cattlemen that their weaning weights are too high. Everyone knows that heavy weaning weights are necessary for profitability, but just to be contrary, let's look at the whole picture.

There are four ways to wean heavier calves:
- creep feed them while they are on the cows
- keep them on the cows longer
- have them born earlier
- use big heavy milking cows bred to big bulls

The last two choices have become common in the last thirty or forty years; people still talk about a thousand-pound cow as being normal, but today, most commercial beef cows weigh 20–40 percent more than that, and most "spring calving cows" calve closer to January than to May.

It is not hard to produce heavy weaning calves; all of the four practices above will raise the average weaning weight in a herd, but all of these practices come with a cost. Creep-fed calves will be heavier but will incur a cost for feed and labor; they will also be fatter and normally will sell for less per pound. Calves can be made heavier by leaving them on the cows longer; whether this makes financial sense depends on what it costs in reduced number of cows carried and the cost of keeping the cow in milk longer. With the right kind of moderate- to low-milk–producing cows able to get by on grass alone, this can be a viable option; this kind of cow used to be the norm but is rare today.

Big cows have bigger calves than do small cows; they also eat more, wean a smaller percentage of their body weight in calf, and wean fewer calves due to calving difficulties and poorer breed back percentages. The calves that survive being born in the winter generally weigh more than calves born later, but fewer survive to weaning, and they cost a lot more in feed costs and poor breed back to produce. If having heavy weaning calves means that you produce less beef per acre that costs more to produce and brings less per pound, tell me again why we have to have heavy weaning weights. Chasing productivity for productivity's sake is not good business for the producer; it is very good business, however, for the people who sell stuff to increase productivity.

The secret to good management in any field is threefold:

- a clear sense of what is to be achieved
- a well-thought-out, comprehensive plan to achieve these goals

- a way to monitor the results of the plan and make the adjustments that are sure to be needed

For the last forty years or so, most agricultural research and education has been geared toward increasing production, with the assumption that if production is high enough, it must be profitable. If profit is required, the plan must be explicit on how the profit is to be made. Even if you can see with your plan how to make a profit, this does not guarantee profit. It is, however, an almost sure thing that you will *not* make a profit if you can't see one in the plan. Most farmers and ranchers hate desk time, but the most important job of any manager is planning. The important fact is not how much money is coming in; what is important is the difference between what is coming in and what is going out.

It is good management to learn how to do something cheaper; it is better management to learn how to not need to do it at all.

In the early 1970s, we were "doing it right," and we were producing a lot of beef. Our carrying capacities were high, our weaning weights and weaning percentages were high, and we were losing money most years. Everything we were doing was recommended practice and would cash flow in a normal year, but we were spending more money than the potential for profit would justify. The cattle market wreck of 1973–74 was a wakeup call for us; it was close, but we managed to survive and move forward to put together a ranch that was both productive and profitable. When we became profitable, our weaning weights were more than a hundred pounds lighter than they had been at their highest, and ranching was a whole lot more fun.

CHAPTER 9

Work with Rather than Against Nature

Humans are by nature impatient and straightforward; when we encounter difficulty in achieving whatever we are attempting, our response is likely to be "Damn the guns; full speed ahead!" Many of us in agriculture grew up with the frequently reinforced admonition that if we are not sweating, we are not working; it follows, then, for those of us raised this way, that when we have a problem, we have to "do something." The hardest thing for many ranchers to accomplish is to sit quietly and think about the situation; it is even harder to plan things out on paper, because for many it is a disgrace to be caught in the house after daylight, and we are too tired to do it at night after a full day of digging post holes and other executive level high-value activities. We want to jump in with both feet and solve the problem, kill the weed, plow the field, feed some supplement, do something; the problem is, attempts to solve production problems with technology quite frequently have the long-term effect of making the problem worse. Using high rates of acid salt fertilizers destroys the soil life that nature depends on to provide the steady stream of available nutrients for the entire system. Poisoning animal or crop insect pests can also kill a wide range of beneficial insects and cause even worse pest outbreaks. Nitrogen applied at high rates can force plant growth to exceed the ability of the soil to give up nutrients, thus creating mineral-deficient plants that are prone to attack by disease and parasites. The chemical imbalance of these plants makes them less than ideal forage for grazing animals, and their use can cause problems with animal health or production. Research in Argentina has shown that fertilizing grasses with even moderate rates of nitrogen lowers the rate of gain for fattening animals by changing the protein-to-energy ratio. Removing all predators allows the numbers of the prey species present to explode to pest levels. The government predator-poisoning programs of the 1950s allowed jackrabbit and packrat populations to increase to the point that some areas were overstocked just with the rats and rabbits.

Pest organisms, be they weeds, jackrabbits, grasshoppers, or mesquite trees, become problems only when their numbers increase dramatically beyond normal population levels.

In most cases, this occurs because the local environment consisting of soil life, vegetation, and animal life becomes less complex. Biodiversity has a bad name with some farmers and ranchers, but biodiversity is agriculture's best friend. When biodiversity is high, no one species has numbers high enough for that species to become a pest. If biodiversity is low, a succession of pest organism explosions will likely occur as nature attempts to fill the unused ecological niches. Weed and brush encroachment is nature's way of covering bare ground and utilizing the sunlight, moisture, and nutrients that would otherwise be wasted to the system when we trash the existing vegetation with poor management. Attempting to kill the weeds with poisons, mowing, or tillage will further simplify the local environment and set the stage for outbreaks of new weeds. Like cold viruses on a schoolyard, weed seed is always there; weeds become thick enough to be pests because they are better adapted to the local degraded conditions then are the desired forage plants. Changing the conditions so that this advantage no longer exists solves weed and brush problems.

This is a good place to hit a lick at one of my pet peeves. As I have said before, there is no such thing as an invasive species either plant or animal; there are only species that are adapted to certain sets of environmental conditions. If you don't like what is growing, change the long-term growing conditions by managing the water cycle, mineral cycles, and energy flow. Planned grazing can bring about this change by stopping both overgrazing and over-rest while improving the functions of the water cycle, nutrient cycle, and energy flow. All parts of the soil-plant-animal complex are bound together by a multitude of interrelationships; actions applied to any part of the complex have far-reaching effects that are often hard to predict. Any practice that is detrimental to the health of any part of the soil-plant-animal complex will, in the long term, be damaging to the entire complex. There are no quick and simple answers when dealing with nature, and the effects of management practices, particularly those that destroy life, should be closely monitored to ensure that their total effects are what are desired.

It is getting harder to find even little scraps of "the natural world"; it might be useful to think about natural tendencies before man gets involved. In nature, extremes are always at a disadvantage; the biggest bull elk may win a lot of his fights but is also the one most likely to winter kill when things get tough. The bison cow that gives a lot of milk is apt to lose too much body condition to rebreed or perhaps even to survive. If animals become so numerous that they foul their environment, they will be cut back in numbers by disease, parasites, and predators. There are no monocultures in nature; plants concentrate in the areas best suited to their needs but always in mixtures, and the proportions of the various plants in the mixtures ebb and flow with changing conditions.

Nature uses those plants and animals that have the best chance to succeed in a given environment.
If you have a lot of weeds, nature is telling you what she thinks of your management.

If nature were a person, she would look at an area and say, "Here is what we have to work with; let's see what wants to grow here." We, on the other hand, say, "This is what I want to grow here, and this is the way I want to grow it. What must I do to make it work?" Only rarely do we even consider that the most logical and profitable course of action might be to learn how to utilize what nature provides in a nondestructive way. There was a time, and still is in isolated instances, when good farmers maintained the productivity of their soils and controlled weed and disease problems with such things as crop rotations, cover crops, including animals in their production program, and following a nothing-wasted program. No system of agriculture has ever survived long term without animals to utilize the crop residues, recycle mineral nutrients, give monetary value to soil improving crops, and maintain soil health.

A note on weed and brush control: most weed problems are the direct result of poor management and can be rectified simply by changing the management to favor the desired forage plants. In some cases, the "weeds" are not weeds at all but rather are the plants that are best suited to the local environment. For example, in soils that are severely deficient in phosphorus, shrubs (brush) and tap-rooted forbs (weeds) are normally more successful than grasses; these plants are more likely to be infected with phosphorus gathering mycorrhizal fungi and their deep roots give them the ability to gather phosphorus from soil strata that the grasses cannot reach. Killing the brush and weeds on such sites to reduce their competition with the grass will only result in even less grass production when the shrubs are no longer there to bring phosphorus into the system. We are able to do a great many things with technology, but just because we can do it does not necessarily make it the right thing to do; look long and hard at the strengths and weaknesses of your local environment—this includes human knowledge and abilities as well as financial conditions and markets—and decide which set of production practices makes the most sense. The closer the operation fits the local conditions, the fewer inputs will be required, and the better the chance that the operation will be profitable.

A big part of our current economic and ecological mess is due to political and economic distortions that have little or no basis in logic; it started with the "plant fencerow to fencerow; get big or get out" advice from the Washington DC mental midgets, which has gotten progressively worse as more and more people buy into the concept of agriculture being an industrial rather than a biological endeavor. The ethanol-from-corn program is an excellent example of an agricultural policy designed by special interests groups, sold by bribery, and rammed through by vote-buying politicians. I cannot comment intelligently on the actual costs and efficiencies (I have read too many conflicting accounts), but it is obvious that we have distorted the grain markets worldwide and put a lot of money at risk based on the thin hypothesis that oil and gas are "bad" and ethanol is "good." Science does not support this position, which is totally political in nature, and so far, neither does the market; I do not want to risk my financial well-being on the whims of politicians who are trying to placate everyone involved.

A second good example of muddy thinking that has gotten people in trouble is the command that we produce cattle that "fit the box" or "meet the demands of the trade." The packers have absolutely no interest in whether anyone else makes a profit; their actions are driven totally by self-interest—which, incidentally, is exactly the way it should be. By the same token, we have the responsibility to look after our own interests. *The only logical criteria for selecting animals is whether they can be produced profitably under your conditions.*

Understand that the feedlot industry has nothing to do with the cattle business. It is solely concerned with selling cheap grain and renting lot space.

The packers and feeders don't want a ruminant capable of thriving on grass. They want a pig in a cow suit that can eat thirty pounds of corn a day, gain five pounds a day, and dress 70 percent.

Most of the money spent fighting the "problems" of weeds, diseases, parasites, and predators, as well as a lot of the money spent for supplemental feed, can be saved by making changes in management practices. If you have ever calved out a set of heifers in January, you probably will agree that there are valid reasons that baby deer, elk, and even packrats are not born in the wintertime. In chasing after heavy weaning weights, we have adopted practices that are frequently hard to justify either economically or managerially. Calving months before new forage is available and breeding oversize cattle that are unable to meet their nutritional needs by grazing what grows in the pasture are two such practices. If the purpose of your operation is to have bragging rights on weaning weights at the coffee shop, it may make sense to breed heavy milking fifteen-hundred-pound cows to calve in January but I suggest that you not try to compete in the cost to produce a pound of calf contest. If your goal, however, is to run a profitable ranch and be able to pass it on to your kids, I would suggest that you consider calving moderate to small low-milking cows in late spring or early summer; it would help also if these cows had the ability to get fat on decent pasture even while suckling a calf. You probably will have to hunt to find cows like this; in an effort to please the packers, we have just about bred the ability to fatten on grass out of our cattle. Let me see if I can justify this wild-eyed bit of heresy. Oklahoma State has done some very good work showing that the most important trait in profitability for beef cows is efficiency of reproduction, and the most important factor in determining how quickly a cow rebreeds after calving is how fat she is the day she calves; the simple and cheap way to have fat cows at calving is to set the breeding date so that cows have thirty days or so of unlimited green pasture before they calve. Dr. Jan Bonsma reported that cattle giving birth on or close to the longest day of the year (June 21 +/-) breed back much quicker than do cattle calving on days with shorter lengths of daylight; as a bonus, heifer calves born during long daylight periods come into puberty earlier and have greater lifetime fertility than do heifers born when days are short.

Several economic studies on beef cow operations suggest that the average producer spends something like 70 percent of his or her annual budget on winter feed; it takes a lot more feed to winter a wet cow than it does to winter a dry cow. If we know that winter-born calves suffer more death loss and sickness (I once lost twenty newborn calves in one night of a January ice storm), and if we know that winter calving cows are less fertile and cost a lot more to feed, tell me again why we don't calve on spring grass. To go a step farther in promoting this heresy, New Mexico State has shown that low-milking cows require something like 10 percent less feed than do high-milking cows of the same weight; you can run 10 percent more low-milking cows on a given amount of grass than you could if they were all high milking. They further found that after sixty days, the calves all gained at about the same rate regardless of whether their dams were high or low milking. No serious person is going to argue that small, low-milking cows bred to calve late in the season will wean bigger calves than big, heavy milking cows bred to calve in the winter, but the former scenario will nearly always be more profitable than the latter.

Look over the data below that was collected over several years from a large number of cow-calf operations in the southeast United States. The data has been extrapolated to a 400-acre farm that is capable of producing enough forage to run a hundred cows weighing a thousand pounds each. You can run more small cows than you can big heavy milking cows on the same amount of grass, and the smaller cows will both be more fertile and wean a larger percentage of their body weight in calf weight; in addition, the smaller calves will bring more per pound if sold at weaning, or they will be cheaper to feed over winter if they are carried over to graze the next season.

Southeast Integrated Resource Management Data

No. Cows	% Calf crop	No. Calves	Cow Weight	Ave Calf wt	WW/Cow exposed	$/cwt of calf	Total Value
100	87	87	1000	510	444	$102	$45,257
91	85	77	1100	530	448	$101	$41,218
84	84	71	1200	580	490	$99	$40,768
76	80	61	1300	600	482	$97	$35,502
71	79	56	1400	610	481	$95	$32,452
67	77	52	1500	610	473	$95	$30,134

It used to be common practice in some range areas to leave calves on the cows until the following spring; most of these cows were not giving much more than a sip of milk a day by nine or ten months into lactation, so it was not a serious drain on their resources—but that little bit of milk gave the calf a great start. Some researchers believe that calves that have access to even a small amount of milk through about ten months of age have a significant advantage in the ability to store energy as fat and in becoming efficient ruminants. The heifers born under this regime breed earlier and are more fertile long term, while the steer cattle fatten more easily.

The cows would kick the big calves off when new grass arrived and before the new calves were born; this scenario is still the norm among wild grazing animals. This program is a compelling one on many counts with the right kind of genetics; the heifers develop well and learn how to be cows by watching as their dams give birth and care for the new calves. When we began to select for heavier weaning weights (more milk and larger mature size), this program was no longer feasible, as the larger amounts of milk placed too heavy a burden on the cows.

For many years in west Texas, the sheepmen waged a never-ending war with eagles that wintered in the area and were serious predators of new baby lambs and goats; it was not unusual for people to lose 25 percent of their lamb crops to eagles. When it became illegal to kill eagles, people were forced to move their lambing season farther into the spring to when the eagles had all returned to their summer ranges. The results surprised a lot of people; no more loss of lambs to eagles, and fewer problems with coyotes and other predators. The later-born lambs were lighter, but there were more of them due to less predation, and they brought more per pound. Just moving the lambing season to later into the spring could easily make a difference of two thousand dollars or more per hundred ewes, which at that time was a lot of money.

When I got back into the sheep business on the Red River ranch, I made two major changes; I used guard dogs to protect the sheep, and I forbid anyone on the ranch to kill a coyote. This caused some of my sheepherder friends to look at me sort of funny; one day Clay Mitchell, whose name strikes fear in the councils of the coyote, bobcat, and lion, was riding around the place with me when a coyote walked across the road in front of us. Clay, thinking I had not seen it, punched me in the ribs and said, "Walt, there is a coyote." When I didn't grab the rifle on the seat between us, he said again, "There is a coyote!" I looked and said, "Yeah, his name is Tony." We drove off and left Tony sitting beside the road with a tongue-lolling grin on his face and Clay looking closely at me to see if I had brains leaking out. My rationale was that the coyotes that were on the ranch were not sheep killers; many generations of coyotes had come and gone since the goat fiasco, and as coyotes are highly territorial, my coyotes that were not sheep killers served as guardians against foreign coyotes that might be less civilized. I almost broke my own rule when I saw a female coyote with a crippled front leg; crippled coyotes often become stock killers, and I was tempted to shoot her. I didn't shoot her, however, and I saw her regularly for the next four or five years; she raised pups every year in the same patch of timber, and to my knowledge, neither she nor her pups ever killed stock.

We did have one bad actor that wasn't with the program, so I shot the fat S.O.B. while he was eating a freshly killed lamb; we lost very few lambs over the next few years to predators, and none that I know of to coyotes. A neighbor told me a similar story; he had a big dog coyote that he saw most mornings as he made his rounds, but neither it nor other coyotes ever bothered his sheep. He hired a young man to feed for him while he and his wife made a trip, and the very first morning, the young man killed the coyote. They began to lose lambs to coyotes almost immediately and were still having trouble the last time I asked.

CHAPTER 10

Stockmanship

This is number 6 in a series of factors that contribute to the profitability and stability of grazing operations; it probably should be first on the list. All disease and parasite problems are worse in stressed animals, and many of these problems can be eliminated or greatly reduced by simply using good husbandry. Usually, disease and parasitism problems are really just symptoms of deeper problems and weaknesses of management.

The main job of the stockman should be to ensure that all of the needs of his animals, both physiological and psychological, are met all of the time.

Before you scoff at me for worrying about the psyche of a cow, note that stress-free animals produce, and they don't get sick. Several years ago, we were buying stocker cattle, and every time the order buyer delivered a load of calves, he asked how many we had lost; all of his clients were losing cattle. It was one of those years when respiratory disease was rampant, and everyone had sick cattle. Each time he asked, I was able to tell him that so far, we had not lost an animal and had treated very few. I didn't try to explain this to him, but the difference was that we had a young man, trained by Bud Williams, who worked constantly with the new calves to relieve stress and make them content with their new situation. No one will believe the effect that this can have until they see it happen. I know of good stocker operators who, after years of high death loss and high medicine bills, have cut their mortality rates to almost nothing and routinely expect new calves to start gaining immediately; the only change they have made is in the way the new arrivals are handled to relieve stress. Any dairyman will tell you that a rough or careless milk hand will reduce milk flow by upsetting the cows; stress reduces animal performance and causes illness.

The goal of the stockman should be to make the animals in his charge as content and happy as possible at all times.

To put it another way: Elsie, the contented cow, really does give more milk. Bud is once again teaching stockmanship schools, and I would urge anyone serious about livestock production to contact him at www.stockmanship.com.

Proper handling is particularly important during stressful procedures such as weaning or shipping; accomplishing what needs done in a nonstressful manner can easily be the difference between profit and loss. Having the animals at the right place at the right time means that whatever needs to be done can proceed quickly without a lot of turmoil or wasted effort for man or beast. After years of riding into a pasture at daylight to see wild cattle disappearing over the ridge on the back side, penning cattle or sheep that are being moved regularly to fresh pasture is a matter of fixing the gates and getting out of the way. Planning can play a large role in stress reduction by ensuring that animals can be moved with a minimum of stress into working facilities, into shelter from weather, or into the shipping pens. All animals are creatures of habit; they like the familiar and distrust the unfamiliar, and any radical change in their routine is stressful. If things that must be done, such as penning for calf working or for shipping, can be made to resemble the routine paddocks shifts that they enjoy, cattle will remain calm, and stress levels will stay low.

To be successful, stockmen need to learn to understand exactly how their actions affect the animals; they must learn when to apply pressure and when to relieve it. Applying pressure (entering the animals' comfort zone) causes the animal to move, while relieving pressure allows the animal to relax. I am not going to attempt a lesson in stockmanship in these pages; this is something that must be learned hands on. However, it starts with learning to watch the animals and relating their reactions to your actions. I would urge anyone who is involved with livestock to contact Bud Williams at www.stockmanship.com or Tina Williams and Richard McConnel at www.handinhandlivestocksolutions.com or Joel Ham at Big Lake Texas, or Guy Glosson at Snyder Texas; all of these people can teach this most valuable of skills.

Stock that is moved regularly to fresh pasture is not stressed by being moved and can be handled and penned with very little labor. Planned weaning on good pasture, where calves are able to see and smell their mothers and touch noses through a fence can drop sickness and reverse weight loss at weaning. Adding a few dry cows or big heifers to the weanlings has a calming influence (you wouldn't put a bunch of junior high kids together without chaperones) and makes it easier to start the group on their rotation moves.

I am ashamed to admit that I weaned for years in a board fence dry lot weaning pen, with all of the dust, confusion, and stress-caused sickness that this entails. I finally changed when Steve McCune talked me into weaning on an electric fence in the pasture. The first time I tried this, it was late in the evening by the time we got the calves separated from the cows and each group turned into its respective paddocks. We had made sure that we had good pasture available, and both groups dropped their heads and began grazing as soon as they were turned out; there was a little talk back and forth across the fence, but nobody was upset. It was just the "Where are you?" and "I am over here" kind of talk. As it started to get dark, the talk picked up a little as more pairs hooked up along the fence, but there was still no urgency and no sign of frightened or upset animals in either group. I lived a couple of hundred yards from the working pens and was normally serenaded all night for three or four nights at weaning time. This time, though, just as I finished my supper, it got deathly quiet; there was no bawling, no mooing, nothing. I told my wife, "That's it; the fence is down and they are back together. I will start over in the morning." At first light, I was at the weaning paddocks to see all of the cattle still lying down; cows on one side of the fence and calves on the other side, and all of them sound asleep. If anyone had tried to tell me this happened before I saw it, I probably would have thought he was a liar even if I didn't say it. The cattle were calm, because they were not in stress; the cow could smell her calf right across the fence, even lick his face, and she knew that junior was ok. Because momma wasn't worried, junior wasn't worried. Junior sure would have liked to have a little milk for breakfast, but his belly was full of good grass, and he wasn't hungry; momma's bag was tight, and she wished junior would nurse to take the strut off, but neither animal was in stress.

Normally, we open another paddock for the cows after two or three days, and a few days later, we start them back on their normal rotation: the calves, with their two or three "baby sitters" are usually ready to start on their own rotation in about a week. No preweaning shots, no medicated receiving ration, no sick calves, and I am able to sleep all night.

We plan to have quality forage available for the calves in their weaning paddock and plenty of feed in the adjacent paddock for the cows; I don't like to have feed trucks coming and going, and I want the calves to eat the same diet, less milk, that they had been on with their mothers. I have weaned thousands of calves on the wire since that first batch, and I only remember one calf getting seriously sick during weaning. This calf got out of the weaning trap but didn't get back with the cows, and when I found him he was franticly trying to get back with one group or the other. He had evidently gotten out during the night, and by late afternoon when I got him up, he was in full-blown "shipping fever."

When working stock, go slow to get through quick.

Whips, half-trained biting dogs, and chousing of any kind have no place on a run-for-profit ranch, but it is usually easier to train stock and dogs than it is to train cowboys. Driving stock too fast stresses animals because of exertion but also because it forces low-ranking animals to intrude in the space of higher-ranking animals and thus upset the social order. A cow that has been pushed into the space of a higher ranked cow feels a lot like you would feel if you found yourself at a funeral in your underwear. The normal walking gait of cattle is about three miles per hour. Horses walk about four miles an hour, so herders on horseback must constantly guard against crowding and against following directly behind an individual animal. All livestock are prey animals and as such are very sensitive to having predators (people, dogs, saber tooth tigers) following directly behind them in their blind spot. Whoever is bringing up the drag should work back and forth across the herd and move no faster than the slowest animal.

The wreck is never the livestock's fault; cattle (or sheep or goats or whatever) only do what cattle do.

If what they do is not what you want, then it is you who needs to change, not them. Along that same line, animals always tell you what they are going to do if you pay attention; the herd does not "suddenly blow up." There are reasons for wrecks, and they are always apparent to someone who knows how to pay attention. As a reformed—or maybe just a worn-out, sure enough, hell-bent-for-leather, brush-popping—cowboy, I can say, "Show me a bunch of wild cattle, and I will show you a bunch of wild cowboys." The good stockman will watch his animals and modify his own behavior to ensure that his charges are content and that what needs to be accomplished gets done without stress to man or beast.

Stockmanship is more than just getting animals to behave in a certain manner; it also entails a lot of what used to be called animal husbandry. I was shocked recently to discover husbandry defined in a dictionary as an archaic term; in my mind, husbandry is still the process and the act of caring for all of that over which one has responsibility. The husband cares for and protects the wife and family, and the stockman cares for animals in exactly the same manner; if that statement offends anyone, I feel sorry for them. I contend that we lost a lot when we changed the name of the course of study from animal husbandry to animal science; animals are individuals that require more than a scientifically balanced ration and the proper kind of immunization program. When we got penicillin and DDT, it looked for a while as though we had finally conquered animal diseases and parasites; what we got instead was the ability to keep animals alive in inherently unhealthy situations.

Not long ago, I visited with a young man who had been operating broiler houses for about eight years; he had early-stage emphysema from breathing the foul air of the chicken houses, and yet he could not afford to quit and do something else. He was in debt, with everything he owned tied up in the chicken houses and with a young family to look after. He would not allow his wife or children to come into the houses because of the health hazards, and he was worried that he would soon not be around to care for them; that is not the way that a farmer should have to live. We have taken a biological endeavor, agriculture, which was a great way to live and to raise kids, and have turned it into an industrial process with all of the downsides but few of the benefits of industry. In the late 1960s, I got sick and finally went to a clinic, where I gained a lot of sympathy for cattle being pushed through the chutes and being jabbed from all sides. One doctor asked what chemicals I had been exposed to in the last year; I picked up a notepad from his desk and made a list of eighteen or twenty chemicals that I had used recently and handed it to him. The doctor read over the list and pitched it back to me, saying, "If you are going to use these materials, I can't help you." It turned out that I had not been poisoned, I had brucellosis. It was treated, and I recovered, but the doctor's comment was a wakeup call. We have become too dependent on chemicals, antibiotics, and immunizations, even though hygiene and stockmanship can often correct the problem rather than just treat the symptom.

I have made that statement to enough groups to know that a percentage of the people who read this will immediately think, "Yeah, maybe, but how about_____?" We cannot cure every ill with stockmanship, but we can greatly reduce both the incident and the severity of animal diseases and parasitism with a little thought. Moving animals regularly to clean pasture and water can do wonders; you would not feed your child from the same plate meal after meal without first washing it. Why would you force your animals to graze among their own waste and drink nasty water? Disease organisms and parasites are always present; they become problematic only when they are present in high enough concentrations to overwhelm the animals' immune systems or when those immune systems are weakened.

Ten germs won't make you sick;
Ten million will.
The solution to pollution is dilution.

Many of the parasites that affect our animals either hatch or spend a part of their lifecycle in manure; horn flies, face flies, and several kinds of internal parasites come to mind. The reaction of modern agriculture to this fact has been to find ways to sterilize the dung. I have a letter in my files from a government scientist saying that I should not worry about the ending of the USDA program to import foreign species of dung beetles into the United States, as dung beetles would never be able to eradicate horn flies; besides, he said, "We now have products that we can feed to cattle that will ensure that nothing will ever be able to live in their manure." I think he must be kin to the fellow who wrote the report saying that manure had so little value that it was not worth spreading on the fields. I am pretty sure that it was a cousin of both of these guys who wrote in the USDA yearbook of agriculture that if dung beetles are bothering you, DDT will kill them.

If animals are moved away from their own body wastes regularly, many of the ills associated with dung-dwelling pests are greatly reduced; if the health of the life in the soil has been promoted with high stock density grazing and no use of poisons, the reduction is even greater. In this case, the dung is used as a feed source for multitudes of beneficial creatures and is no longer a habitat for pests.

Contaminated water is a prime source of disease, and it is well worth the effort to reduce the amount of dung and urine entering the water supply; piped water is easy to keep clean, but water quality can be increased even in ponds or dirt tanks. Animals should not be allowed to wade into the water source but should be confined to drinking from drink tanks below ponds or from areas fenced to allow room to drink but not room to wade. If lactating cows are allowed to wade into the water source, calf scours are sure to follow.

A factor that deserves serious thought when designing a fencing plan is how to keep areas that are in frequent or constant use to a minimum. Lanes, barn lots, and water points that have animals present for prolonged periods all become sources of parasites and diseases. If it is feasible, including multiple water points in a group of paddocks grazed together can both reduce spot degradation of the soil and forage and help reduce parasite infestations and incidents of disease. The location of gates can become a stress factor if the locations are such that they interfere with the instincts of normal animal behavior; if moving animals through a particular gate results in a fight, with animals cutting back and calves running through fences, you need to look at the location of the gate from the animals' standpoint and fix the problem. A final thought on stockmanship: if you are in a bad mood or upset, the animals will know it and will react with distrust. If you are grouchy and mad, it would be better for all involved that you just go to the house and have your mad fit all by yourself.

If you do not like and respect animals, you will be a lot happier selling shoes, and so will your animals

---

Stockmanship is more than just getting animals to do what you want;

It is the ability to see the situation from both your own logical perspective ("That ramp is steep, but they can make it without slipping down") and from the animals' viewpoint ("That is scary, and I don't want to go there.")

CHAPTER 11
Correct Stocking Rate

Stocking rate, the number and weight of grazing animals present on an area for the grazing season, is a major factor influencing profitability and stability. Stocking rate is a direct quantification of forage demand and affects all aspects of an operation. If stocking rate is too high for a prolonged period, nothing else works as it should, and an unstable situation is created. Much of the money spent on feed supplements and on feed substitutions like hay is due to stocking rates that are too high for the circumstances. Having more animals than the grass can support is a guaranteed profit killer and a major reason so many ranches are not profitable.

One of the most valuable traits of ruminate animals is their ability to cheaply gather low-value forages from out in the rocks and brush and convert them into high-value food protein and energy. No forage harvesting system devised by man can come close in cost or ease of management to the properly managed grazing animal. To be profitable, a grazing operation must maximize the number of days the animals gather their own feed by grazing. Wild ruminants, and livestock before we bred the ability to become fat out of them, were designed to put fat on their backs when grazing is good and then utilize this fat when grazing is sparse. By selecting for animals that fatten easily and making sure that they have forage available to turn into fat, it is possible to winter cattle very cheaply. It takes the right kind of animals but it also means that you have to have more grass than cattle. Growing up, the only hay I ever fed was to saddle horses or other stock that we left up in the pens for some reason; we were able to do this because Dad did not overstock his country. Back in the 1940s my father ran straight-bred Hereford cows; on decent pasture, those old girls would get fat by midsummer and still bring a decent calf to the weaning pen that fall. These were not big cows; most of them probably weighed closer to 900 pounds than to 1000 pounds, but at nine to ten months, their calves would wean at around 450 pounds. A cow that failed to regain condition before frost was considered a "hard doer" and likely would not be around long. If a cow didn't breed or lost her calf, by fall she would be town dog fat and ready to make some pretty good beef.

Planned and time-controlled grazing is a powerful tool for achieving profitability and stability, but no management program can overcome incorrect stocking rates. Overstocking brings about reduced animal performance, loss of range health, and increased financial and ecological risk. Understocking is inefficient as well; it reduces animal production per unit of overhead, thus increasing cost of production; additionally, if prolonged, it will bring shifts from grassland toward woody plants.

> When a cow is grazing in the pasture, she is working for you. When you are feeding a cow, you are working for her.

I know! I know!

It is important to understand the difference between overgrazing and overstocking. Overgrazing occurs only to plants and on a plant-by-plant basis. A plant is overgrazed only if grazed while it is growing on stored energy because it has not had sufficient quality growing time to recover from the previous defoliation. Overstocking refers to the effects on an area rather than on a given plant. Overstocking is one cause of overgrazing and is a common and serious management fault, but reducing stocking rates will not eliminate overgrazing. When an area is continuously exposed to grazing animals, some plants are damaged by overgrazing while others are damaged by over-resting. This occurs because when a large area with a lot of plants is available to the grazing animals, not all plants will be bitten uniformly. Those that are bitten will regrow fresh leaves that are more attractive to grazing animals than the older and tougher leaves of plants that were not bitten. This causes some plants to be bitten and rebitten until they are damaged by overgrazing, while the unbitten plants accumulate large areas of old, ineffective leaf, which shades their growing points and retards growth. This phenomenon occurs whether stocking rates are high or low and is responsible for much of the range degradation that has occurred worldwide. Many of the most valuable forage plants are at least as susceptible to damage from over-rest as to damage from overgrazing. Grassland is always degraded when stocking rate is too high, but the effects are especially severe when animals are provided with substitute feeds during the growing season rather than being removed from land that can no longer feed them.

If you routinely feed hay or other energy feeds during the growing season or early in the dormant season when there is no snow cover, it is a pretty good bet that stocking rate is too high and that profitability could be increased by reducing stocking rate. If your cattle will not perform without the additional energy, you have a serious genetic problem as well as a management problem. Aside from overstocking, the most common causes of this problem are poor matching of nutritional demand with forage supply (i.e., calving out of season), cattle that have been genetically selected to perform on grain rather than on grass, poor grazing management, or poor pasture. Ruminants have a valuable trait: they can store excess energy as fat when times are good and maintain health when feed is short by drawing on this storehouse. To do this, however, they must have access to excess feed during some period of the year. This excess is not "wasted grass"; it is what can keep the rancher out of the feed sack and hay barn. If you must have enough animals to use all of the spring flush, you had better do it with animals that are not going to be around in January or even in August. All cattlemen love to feed cattle; get your stocking rate in line with pasture production, and get the same enjoyment by watching the excitement when you open the gate to a fresh pasture break.

CHAPTER 12

Correct Stocking Mix

Areas that produce forbs and browse as well as grass can often benefit from being stocked with more than one species of grazing animals. Brush is eaten by goats, and weeds are eaten by sheep, which increases the amount of nutrients in the mineral cycle and the rate of their cycling. Bringing this material into the food chain, where it is bitten and regrows, raises the amount of energy flowing through the system by causing more of the total vegetation to be in a growing vegetative stage and thus more effective in capturing and converting solar energy to biological energy. The weeds and brush become an asset instead of a liability and generate income rather than causing expense. When these plants are eaten, the minerals they contain enter the nutrient cycle much faster and more completely than if the plants become senescent with age and fall to the ground, where biological decomposition can occur only at the end of the growing season. All factors that increase the amount of minerals in the nutrient cycle or the rate with which minerals cycle increase the amount of life within the system.

Bringing shrubs and tap-rooted forbs into the feed bank can be especially valuable, as many of these plants are more effective than grasses in taking up some minerals and can tap soil strata not reached by grasses. In some mineral-deficient soils, shrubs and forbs play a significant role in making minerals available to grazing animals. In these areas, cattle tend to learn to eat more browse, but this is a learned behavior and may take several generations; adding some goats to the stocking mix can speed up the process of putting minerals that are locked away in brush into the nutrient cycle.

Long-term productivity and stability come from having all available resources being fully utilized without any being overutilized.

Yes, I know!

The epitome of multiple-species stocking occurs in some areas of Africa where dozens of different species of grazing animals have evolved to take advantage of all of the different types of forage. The giraffe takes the topmost leaves out of fifteen-feet-tall acacia trees, while the tiny dik dik antelope feeds on the tips of forbs and grasses growing in shaded forest; the Cape buffalo and hippopotamus mass graze grasses and reeds, while some antelope concentrate mainly on forbs, others on browse, and still others mostly on grasses. All of the available vegetation is utilized to some extent, so both energy flow and rate of nutrient cycling is high; the result is that the land supports many times the amount of animal life that it could support stocked with only cattle or sheep.

Before we shipped our mohair and wool markets to China, many ranches in the southwest were stocked with a combination of cattle, sheep, and goats; this was a profitable and sustainable program when the combination was properly managed and the proportions of the stocking mix devoted to each species accurately reflected the forage available on the area. Today, the mohair and wool markets are lacking, but meat goats and hair sheep can fill the ecological niche once held by mohair goats and wool sheep; they can also fill a financial need. A story is told about an Oklahoma cattleman who got talked into adding some goats to his program. He got some goats and spent the next three years bitching and moaning about how much trouble those "stinking damn things" were and finally loaded them all up and sent them to town; when he saw the size of the check, he called his order buyer to see if he could find him some good, young nannies.

MINERAL CONTENT OF GRASS VS "WEED"

	Percent of Dry Matter						PPM					
Element	N	P	S	Ca	Mg	K	Fe	Mn	Cu	Zn	Mo	B	Cl
Big Bluestem	1.00	.20	.07	.23	.16	1.02	54	55	3.6	30	.20	11	776
Prairie Coreopsis	1.17	.15	.12	1.05	.42	1.19	52	117	7.9	65	.37	61	957
Percent Advantage Coreopsis	+17	-5	+5	+82	+26	+17	-4	+62	+54	+54	+46	+82	+19

Adapted from <u>Mineral Nutrition of Plants</u> :Epstein

Multiple-species grazing can be a tremendous help in controlling internal parasites, as parasites tend to be host-species specific; sheep or goat parasites die if ingested by cattle, as do cattle parasites picked up by sheep or goats. In humid areas, having some cattle in the stocking mix is almost a necessity to control parasites in sheep or goats. Healthy animals can fight off parasites whose numbers are not overwhelming; grazing with multiple species is one way to reduce parasite numbers.

Adding species can increase financial health by providing more products to sell, by improving forage production, and by spreading market risk. How valuable it is to add species to the grazing mix depends on how well the available forage fits the needs of new species but also on factors such as costs incurred by fencing needs and predator loss control. Not to be overlooked is the specialized knowledge required to manage the new species. More than one successful cattle producer has learned the hard way that sheep and goats are not "just little cows" and that they have special needs to go along with their special abilities. Internal parasite control is vital with sheep and goats, and failure to realize this is the source of the frequent complaint from cowboys that "all the darn things want is a warm place to lie down and die." Parasites can be controlled by a combination of stockmanship and selection for genetic resistance (with chemical wormers used only rarely if at all) but to accomplish this requires knowledge and attention to detail.

Profitability is enhanced when the forage on offer matches the nutritional needs of the animals grazing; this is true of species but also of classes within a species. Lactating females, fattening animals, and young growing animals have high nutritional demands and should have access to the best quality and quantity of feed available. The duration of this high demand will vary; dairy cows require excellent nutrition as long as they are giving milk, while beef cows, which give less milk, need the high plane of nutrition for the four months or so after calving but can then be scaled back to a more moderate level. Mature nonlactating animals have much lower nutritional requirements and can be used to clean up behind the higher-demand animals, thus improving forage quality on the next rotation.

Ruminants are adapted to storing fat when feed is plentiful and using this fat for energy when feed is scarce; keeping ruminants fat year-round wastes money and is detrimental to the animals' health. Areas with a short period of quality forage are best utilized with animals whose production cycle is compatible with such growth patterns. If most forage is grown in the spring, spring calving cows that can give birth and rebreed during the period of high quality or stockers that are present only when forage is good would be better choices than enterprises such as dairying or grass finishing that require long periods of quality forage. Practices designed to improve the length of the quality forage period, such as planting wheat or other temporary pasture, must be analyzed with regard to the cost-to-benefit ratio of the forage produced. Such a practice may be viable if the forage is used by stocker or dairy animals but be far too expensive for use by dry or even nursing beef cows; there are always exceptions, but the most profitable and trouble-free operations usually entail those management practices tailored to exploit the strengths and minimize the weaknesses of their local environment.

Another factor to be considered when setting stocking mix is the likelihood of drought or other weather-related forage reductions. When drought hits, reducing forage demand by removing animals is often a wise move, and stocking mix can be planned to minimize the financial consequences of de-stocking. Young, growing animals are easier to sell and hold their value better than grown stock, especially in bad times when everyone is selling. In an area prone to drought, it makes sense to make up a portion of the stocking rate with young animals such as stockers that can be moved quickly or not purchased in dry years. The proportion of brood stock to stockers will depend on the likelihood of drought and might range from all registered cows where drought is rare to stocker goats bought only in the year it rains where drought is common.

CHAPTER 13

Drought

Droughts are a regular occurrence for most ranching operations, as ranches are likely to be located in areas of natural grassland—and one of the formative factors for grasslands is erratic moisture availability. Drought is not just dry weather; drought occurs when there is a significant reduction in normal precipitation. A desert area that receives only nine inches of rain is dry, but it is not in drought unless annual precipitation falls well below nine inches; an area that is in a forty-inch rainfall belt and receives twenty inches is in a serious drought. Our home ranch in Nolan County, Texas, was in a twenty-inch rainfall area that was very drought prone; local wags said that the twenty-inch average came about because it would rain sixty inches one year and then skip two years.

Where moisture availability is constant, the vegetation tends to be made up of longer-lived plants (trees); if drought kills most of the local vegetation, grasses and forbs can germinate from seed and reproduce quickly, but trees require much more time to reach maturity. The fantastic ryegrass and clover pastures of England and Ireland did not come into being until the oak forests that were originally there disappeared into ship timbers and charcoal kilns; if humans were removed from these areas, the oak forests would return because of the uniformity of the local moisture patterns. The frequency and severity of droughts vary widely according to location; knowing the probability of drought in the local environment is essential information for formulating management plans. Equally important is recognizing the early signs of impending drought; the sooner drought is recognized, the more effectively its effects can be offset. If drought is recognized as a normal occurrence, and it is, then plans can be made to reduce its impact upon the operation and upon the soil-plant-animal complex on which the operation depends. The National Weather Service keeps detailed weather records for many places in the United States, and an examination of these records for your area is a good place to start in determining the likelihood of drought. There is nothing you can do to keep drought from occurring, but you can do a great deal to reduce its impact.

The effect of drought on grassland is in direct proportion to the health of the grassland; rule 1 in drought-prone areas should be to maintain the plants and soil in the healthiest possible state. This can be done by manipulating the ecological processes as described in chapter 6; when the vegetation in an area is composed mostly of healthy, deep-rooted, perennial forage plants, the soil has a high degree of biological activity, and the ground is covered with plant material both dead and alive, the ability of that area to withstand dry weather is high. Conversely, if the forage sward consists of mostly annuals and overgrazed, short-lived, perennial plants growing on areas with a lot of bare soil, even a slight reduction in precipitation will be devastating to forage production. Build the health of your range, your biological capital, when growing conditions are good so that you can survive the drought that is surely coming. It is much easier to maintain healthy grassland than it is to bring degraded grassland back to health; be ready to do whatever is necessary to keep from degrading your country.

Drought does not destroy grassland; Management during and after drought destroys grassland.

We are all familiar with the effects of drought on land and animals, but we seldom consider that it also has tremendous effects on two other areas: our financial health and our mental health. These two are closely related, but let's look at financial health first. Just as it is essential that biological capital be generated when times are good, it is crucial that a financial cushion be built up when conditions are normal. If it is not possible to create a financial surplus and salt part of it away in a "not rainy day" fund, then the entire operation needs to be re-examined and replanned. If the operation is not profitable under normal conditions, it will certainly not be profitable when drought or other unfavorable conditions occur. This fund is not something that may be needed; it is something that *will* be needed. Until this fund can accumulate enough to fulfill its purpose, an arrangement to have credit available when needed should be made; however, although having credit available is necessary, a line of credit, which is a debt, does not take the place of a financial cushion, which is an income-producing asset. A west Texas rancher with years of experience told me long ago, "We all need a drought once in a while to make us cull the stock like we know we should and to force us to quit spending money based on "we want" rather than on "we need."

A rather cynical note on droughts and government "help": When drought does strike, the last thing most people need is donated feed or low cost loans to buy feed. What happens every time funds are made available for "drought relief" is that a lot of people beat hell out of their country by turning it into a feedlot and wind up deeper in debt with a bunch of thin stock at the bottom of the market.

One of the most economically damaging things that a rancher can do is to hold stock on land that can no longer feed it.

The expense of feed and lost animal production is bad, but the real cost is the damage to the land. Droughts have come and gone for eons without permanently damaging the land, because before fences and stored feed and pumped water, when the land could no longer support grazing animals, the animals either left or died. If you have stock that you are determined to save, send them to where it is still raining or put them in a feedlot.

Get them off the land until the land can once again feed them without being degraded.

Even when animals are being fed all they need and more, they will pick every green leaf as soon as it appears, reduce the ground cover, and do serious damage to both the soil and the vegetation sward.

It is very easy to save the herd
And in the process
Lose the ranch

One of the most debilitating of human emotions is the feeling of helplessness in the face of overwhelming adversity; Elmer Kelton in *The Time It Never Rained* (a great book on a sad subject) paints a vivid picture of the human and financial suffering caused by a major drought. Emotion of this intensity can and does cause people to make poor decisions, and it can bring about physical harm; stress is debilitating to humans just like it is to livestock. A major drought is guaranteed to cause exactly this state of anxiety unless plans have been made ahead of time to offset the effects of drought. Such plans must start well before the drought with the goal of building the health of the local environment, biological capital, and financial health; the most useful tool for building biological capital is planned high stock density grazing. Design, implement, and continually update a formal grazing plan for each unit of the ranch; a part of these grazing plans should be time reserves for drought. These reserves should be in the form of higher residuals of forage left in each paddock after grazing; leaving extra leaf on the plants will speed recovery of the grazed areas and maintain reasonable quality of forage. If an area of forage is set aside as drought reserve, both the quality and the growth rate of the forage in such a set aside decreases dramatically when the plants reach maturity. The infrastructure needed for planned grazing, the fencing and water development, will also be invaluable in coping with drought.

Two things need to be done as soon as it is apparent that a drought has started: reduce the demand for forage, and (as growth rates of forage will be slower) increase the length of time the forage has to recover. The ability to determine whether a drought is imminent requires knowledge of the local weather patterns; the manager needs to know when precipitation normally falls and when grass grows. If the rains do not come at the normal time or fall only sparsely, in most areas, some degree of drought is almost certain. If moisture is low or ineffective during the period of peak forage growth, it is past time to reduce forage demand.

Part of planning for drought includes selecting a stocking mix that fits the local likelihood of drought; in areas where the probability of drought is high, some portion of the stocking rate should consist of animals that can be moved rapidly without incurring financial loss. Young, growing animals can nearly always be sold at a reasonable price, as they can get on a truck and go to where it is still raining; breeding females, especially older females and females with young offspring, sell at a discount in drought conditions. Making up a portion of the stocking rate with stocker animals, preferably home raised, allows the manager to reduce animal numbers quickly with minimal financial pain. In dry years, to lower forage demand, the calves are sold at weaning or shipped to grass; aside from being good drought strategy, carrying at least a portion of the calf crop over to be sold the following summer is a consistently profitable program for many ranchers. A calf at weaning has already incurred about 70 percent of the expense required to produce a seven-hundred-pound yearling. This is because the weaned calf has to pay the cost of maintaining his mother for an entire year. The weight at weaning is expensive, but the gain from weaning to yearling weight can be made quite cheaply under the right program. If cows are calving in the spring flush, this program also has the effect of loading up forage demand (cattle weight) when forage is most abundant and reducing demand when forage is scarce. The calves should be wintered on the best available forage with only enough supplementation to keep them healthy and growing normally; calves wintered in this manner will make quick and profitable gains in the spring flush. The temptation is always to plant some winter green stuff or to pour the feed to the calves to make them "do good"; resist.

It is not hard to make cattle fat in the wintertime, but it is hard to make money making cattle fat in the wintertime.

Before the yearlings are shipped early, the manager should look through the pasture day book—*the one that records in one place when what stock was moved into which paddock, how tall the grass was going in and how tall it was when the herd came out on which date and how much it rained and when it frosted and when 20/5 sluffed her calf and when the bulls when in and when they came out. Even all the other stuff that we used to try to remember or write down on the back of envelopes that blew out when we got out to open the gate; the stuff we need to build and keep up to date a "to be culled" list. I know that you will not forget that you saw 88/3 in standing heat on February 2 or that 16/5 prolapsed, having it down on paper will make it certain that these cattle get on the truck if you are not around.* This is also a time to consider downsizing the cows; if the need is to reduce forage demand, it might make sense to replace some of the older cows and the fourteen-hundred-pound cows with heifers that will be one-thousand-pound cows when grown. Sometimes it takes a drought to jolt us in to doing the right things.

There is another option, viable under the right conditions, if some high-quality feed is available, it may be possible to wean calves early and lower the nutritional demand of the cows. Weaning her calf will reduce a cow's need for energy by approximately 40 percent and her need for protein even more, and it will allow her to maintain herself on lower-quality forage; keep in mind that the calf will have to have access to a high-protein diet to replace his mothers' milk. Regardless of how the sell list is decided on, **when it is time to sell, pull the trigger!**

The table below, which is adapted from *The Holistic Management Handbook*, shows the effects of delaying when it is time to move.

The Timing of Destocking

Conditions: 500 animal units (AU), 60,000 animal unit days (AUD) of forage, 160 days to new grass

Action: 1. Do nothing
Result: Out of feed in 120 days (60,000/500=120)
Action: 2. Wait 40 days
Result: Out of feed in 80 days or will have feed for 333 AU 60,000 −(500 X 40=20,000) = 40,000 40,000/120 days=333 AU
Action: 3. Wait 80 days
Result: Out of feed in 40 days or will have feed to save, 20,000/80= 250 AU 60,000 − (500 X 80=40,000)=20,000
Action: 4. Move 125 AU immediately
Result: Will have feed for remaining 375 AU until new grass, 500-125=375AU, 60,000/375=160 days

In addition to reducing the total demand for forage, it is very important that provisions be made to give the forage plants that are growing slowly due to poor conditions more time to recover. A quick and simple way to accomplish this goal is to combine as many herds as is feasible into one herd; for example, if you have four herds each working through twenty-five paddocks, after combining all four herds, the new single herd will have one hundred paddocks to work through. If all of the paddocks have been stocked conservatively, it is possible that you can increase the recovery periods by a factor of three or four and still maintain short graze periods. If you have been using an average recovery period of 75 days during slow growth, you can now use recovery periods of 150 to 200 days. Using one-day graze periods with one-hundred paddocks and one herd, 1 percent of the area will be being grazed (assuming equally sized paddocks) at any point in time while 99 percent of the total area is resting.

Depending on topography and the amount of forage present, it may be worthwhile to use temporary fence to reduce the size of large paddocks; this will likely work if there are paddocks large enough to require graze periods of five or six days. Long graze periods reduce the efficiency of grazing and reduce animal performance; splitting up a large paddock can often increase the number of AUDs that the paddock can provide simply by reducing the amount of forage spoiled and refused by dunging.

It is usually not necessary to use a back fence when rationing out dormant forage. By increasing stock density the grazing efficiency (the percentage of the forage actually consumed by the animals) can be increased; this will increase the number of AU Ds available and hopefully increase the likelihood of making it through to new grass. It is tempting to feed a little hay or other feed to "stretch the grass"; I have never seen this tactic work. What normally happens is that the animals catch the welfare syndrome and quit grazing to wait for the feed truck, and animal performance falls off. A very little high-quality hay fed daily as a protein supplement can increase utilization of low-quality forage, but do not feed enough high-quality hay on any one day to be a significant part of the animals' diet; if you do, you will risk rumen flora upset, similar to what happens when feeding grain to grazing animals. I would prefer to not feed any energy feed as long as grass is available; when the grass runs out, either move the animals or put them in a feedlot. As with any practice, watch animal performance, and do what is required to keep it in the reasonable range; the operative word here is reasonable. Ruminants are adapted to fluctuating body conditions, but there are limits.

> *Short graze periods and uniformity of diet are the keys to good animal performance. This is especially true when grazing low-quality forage.*

Drought forces us to make choices; sometimes the choice is between bad and worse, but planning can reduce the pain. Develop water in excess of what you expect to need; it hurts to have grass and not be able to use it because the tank is dry. Understand the difference between feeding for supplementation and feeding for substitution; when you feed energy as hay, grain, or whatever, you are substituting money for grass, and you are on shaky financial footing. As the gambler says, "You are behind the game and betting on the come."

Remember that the real damage from drought comes from management that results in a loss of biological capital. On even very well-run ranches, when a major drought finally ends, biological succession will have been pushed backward, and there will be open biological niches; nature will fill these niches with organisms that are capable of existing and reproducing in the new and harsher growing conditions. Expect a plague of weeds along with grasshoppers, armyworms, or whatever pest organisms are common to your area. These population explosions will come about because of the reduced amount of life in the drought-stricken area.

Be careful that your response to the situation doesn't increase the problem. Weeds are nature's way to respond quickly to bare ground. If there is nothing growing but weeds and you kill the weeds, you still have no forage but you now have bare ground as well. Pest organism explosions occur because of a lack of bio-diversity. Be careful that your response doesn't further reduce bio-diversity and make the problem worse.

Do not be in a hurry to re-establish the old number of herds. Keep your stock density high after the drought breaks unless there are compelling reasons to not do so. High stock density is a powerful tool for improving mineral cycle, water cycle, and energy flow, thus moving biological succession forward. It is also the most effective tool to deal with the plagues of pest organisms, as it addresses the root cause of these plagues, which is low biological activity.

If you do not have a monitoring program in place, use the end of the drought as a starting point, and begin monitoring the health of your range. With good management, improvement will be rapid after good conditions return, and this will offer an excellent opportunity to learn how the range heals itself, given the chance. Set up permanent photo points to record the changes, and begin a formal monitoring process that addresses the whole soil-plant-animal complex.

Our management determines the health of our grazing lands. If our practices promote healthy ecological processes (good water cycle, rapid mineral cycling, and strong energy flow) biological succession will advance, and our ranges will become both more productive and more stable. Our management during drought is particularly critical, because the ecological processes are under stress, which magnifies the effects of management mistakes. Natural grasslands are extremely stable due to their complexity and to the health (read high organic content) of their soils. These grasslands evolved in spite of drought over eons of time; when nature was managing the show, if drought became severe, the grazing animals either left or died. The drought eventually ended, and the prairie came back just as it had hundreds of times before. Drought doesn't destroy grassland; our management during drought destroys grassland. If we are going to operate in drought-prone areas, we would be wise to study nature's means of range management. Some of nature's most valuable tools are keeping stock density high, matching the demands for forage to the production of forage (especially during drought), matching recovery time to growing conditions, and never leaving animals on range that has lost its ability to produce the feed they need. Do these things in an economically feasible way and your operation will have a head start in the race for success.

CHAPTER 14

Adapted Animals

Over time, animals adapt to the area and conditions in which they live. In the short term, the adaptations are limited to mainly learned behavior: what is good to eat and what is not, how to find feed, water, and shelter from the weather, and some adaptation of rumen flora to local plants and soils. Over the long term, changes in the local genetic pool take place as those animals with favorable genetic traits live longer and produce more offspring. Wild animals in cold climates are larger than the same species that live in warmer areas, as large body mass conserves body heat. Dark colored coats absorb more radiant heat from the sun, so darker animals have an advantage in cold climates but are at a disadvantage in hot climates. Animals adapt, in one manner or another, to local conditions as diverse as soil fertility, length of day, and numbers and types of parasites present. These factors become of special importance when animals are moved from one climatic region to another.

Animals that are born and raised in an area have advantages over animals brought in from outside. The greater the differences (between the old and the new location) the greater the disadvantage of the transplanted animals. A large part of this is due to animals developing resistance to local parasites and diseases, but other factors also come into play. Radical changes in climate and elevation severely stress animals, with the greatest effects occurring when animals go from cold to hot climates, from low humidity to high, and from high elevation to low. Differences in mineral content of forages also play a role in how soon and how well animals adapt to a new area. In many phosphorus-deficient areas, the local shrubs have considerably higher phosphorus content than the grasses; local animals learn to browse the shrubs and cope fairly well, but animals new to the areas that don't have a history of browsing are at a severe disadvantage. Grazing is hard work, and animals moved into a new area have to learn what to eat when and how to best graze the new pastures. This is particularly true of forages with potentially bad effects, such as legumes, endophyte-infected forages, and plants that contain large amounts of toxins. When it is available, animals can and do learn to add variety to their diets to offset the effects of toxins and thus increase the number of species they can utilize as forage. This learning process takes time, as it depends on feedback both good and bad, and can be an extended process when forbs and browse make up a large part of the available forage. Adding local animals to the herd can help the new animals adjust by providing demonstrations of which forages are edible at various times.

It is more than likely that part of the adaptation process relates to the rumen microorganism populations adapting to local conditions. There are microorganisms that can thrive in every environment that is capable of sustaining life; in streams polluted with cyanide from mining operations, microbes have been found that have evolved the ability to not only to tolerate the poison but even to use it as food. A principle of quantitative genetics is that genetic change is fastest in those species with short generation times. Some rumen microorganisms produce a new generation every twenty minutes; it would be illogical to think that these organisms do not develop genetic traits that make them better adapted to the local conditions found in soil and forage.

Animals that have been bred and selected to thrive under one set of management practices cannot be expected to perform well in a drastically different type of management. This is especially true of animals that are moved to a grazing regime after having been selected for several generations to utilize large amounts of grain in their rations. Not only do these animals not know how to graze effectively, but they will have anatomical conditions that prevent good performance when they must survive by grazing alone. To perform well by grazing, an animal must have large rumen capacity, a wide mouth, moderate body size, and the ability to store large amounts of energy as body fat. All of these traits have been selected against either actively or passively by our feedlot-driven cattle industry. The feedlots do not want a ruminant capable of thriving on grass; they want a large slow maturing animal with a minimum of "waste" in back fat and rumen size. These cattle cannot be profitability produced on pasture alone and are profitable on grain only as long as grain is subsidized.

To make the fastest genetic progress, select only for a limited number of traits that improve profitability.

Part of the challenge in selecting adapted animals consists of understanding what the word adapted means; as I mentioned elsewhere, for years I performance tested and selected my breeding animals based on weaning weights and yearling weights. I got what I was selecting for; I got heavy weaning weights—but I also got big cows that needed a lot of grass, lacked reproductive efficiency, and could not perform without a lot of supplementation. This goes back to the concept of managing toward a set of goals. If profitability is a goal, the animal breeder will do well to select animals that perform well under his local conditions without a lot of inputs.

The best animals are the ones that work for you and make you money.

These are seldom the ones, "demanded by the trade" or that "fit the box." Breeding heavy milking, slow maturing, and large body sized cattle may make sense in farm situations with cheap grains. These animals are not apt to be profitable in a harsh range setting or even in a lush environment where they must graze for their total diet.

*Be careful what you ask for;
You might get it.*

Changing conditions by feeding large amounts of hay or providing other types of welfare to meet the needs of unadapted animals is seldom as profitable as choosing animals that fit the situation. Modifying the local environment to suit the abilities of unadapted animals is nearly always a profit killer. When I finally woke up and realized that what I needed was not heavy production but profitable production, I started to actively *select for traits that improve profitability*. Heavy weaning weights mean big cows that give a lot of milk; this kind of cow is expensive to maintain, so her calf carries a high cost to produce. By actively selecting for smaller cows that give less milk and can get fat while nursing a calf, we were able to produce more pounds of calf per acre at less cost. One of the simplest ways to identify these cattle is to note which cattle calve in the first twenty-two days of the calving period; by and large, these will be the animals that are best suited to your management and your local environment.

The amount of money coming in is not nearly as important as the difference between what is coming in and what is going out.

There are two main ways to change the genetic makeup of a group of animals: you can breed animals that exhibit the desired trait to a high degree (this is the expected progeny-difference approach), or you can remove those animals that do not show the desired traits from the breeding herd. In most herds, ruthless culling of animals that do not perform up to standards is the fastest way to make whole herd progress. Cull out the dead wood and then start selecting for a small number of the most economically important traits. For most of cowherds of today, smaller size, easier fleshing, and lower milk production will be among the things that will increase profitability the fastest.

In graduate school, we were taught that fertility had very low heritability and was therefore hard to improve by selection; I contend that the problem is in how fertility is measured. Fertility is a measure of the total well-being of an animal; it is not a single trait but rather the result of the effects of many traits. Animals that are well adapted to the environment in which they are living will be more fertile than animals that, for whatever reason, are not as well adapted. The simple way to select for fertility is to remove any animal with an obvious fault (prolapsed, bad udder, highly parasitized, crippled, or just hard doing), and then cull any animal that does not bring a live offspring to the weaning. The heritability of fertility may be low, but if you sell all subfertile animals, you soon have a fertile herd. This technique holds true for many traits, from internal parasite resistance, to uterine prolapse, to horn fly resistance, to bad feet.

Clifton Marek of central Texas has a unique culling method for sheep that works like a charm; about once a year after he has shipped lambs, he will call the sheep out of their paddock and take them for a walk down the road. Whoever is leading the sheep sets a pretty quick pace, and Clifton watches as the flock stretches out; when he sees a distinct group of stragglers fall behind, he steps in and separates these sheep to be sold. He will catch the animals that are heavily parasitized or have low-grade respiratory problems, bad feet, infections of any kind, foot rot, and just about any other condition that reduces the animal's vitality; in a couple of hours, he has removed the vast majority of animals that either have problems or will have problems in the near future.

The fastest way to make whole herd progress is to sell the animals that don't fit the program.

CHAPTER 15

Short, Properly Timed Breeding Season

There are reasons that baby deer, elk, bison, rats, and rabbits are not born in the middle of winter. All baby animals are born with a summer coat; nature is too smart to take a baby from a cozy, hundred-degree womb and drop it in a snow bank. If you think this is no big deal, next January go out to the calving trap at day break, take off your coat, your earflap cap, your boots, your two pair of socks, your insulated overalls, your shirt and britches, and your long handles, and roll around in the snow. One time, years ago when I was still dumb enough to be calving in February, Homer Gilbert and I pulled into a pasture of calving cows with a load of hay. It had rained, turning to sleet in the night, and the cattle were bunched waiting at the gate. As we pulled in, a cow lying down on the far side of the pasture raised her head and bawled; we threw the hay off as fast as we could, and I left Homer pulling the wires off the bales and went to check on the down cow. The cow was lying in a drainage ditch with two feet of ice-cold water up to her shoulder and using her back to keep her newborn calf from sliding down the steep muddy bank into the water. I grabbed the calf, pulled it back up the bank, and carried it to a level spot on some grass. The cow stood up when I took the calf and followed us to start licking her baby as soon as I laid it down. She had done the one thing that she could do to save her calf from drowning or freezing to death, and that took reasoning power. I went back and got her a half bale of alfalfa to eat all by herself. The cow was a lot better at her job than I was at mine; she did everything that she could do to protect her baby; I should have been dunked in the ice water for putting her in the position of calving in cold weather with no grass.

Parturition and lactation require large amounts of energy, and nature fills this need by having events with large energy demands occur when feed is plentiful. In a cattle-breeding operation, fertility, measured as the number of live offspring born per female exposed to the male, is the most economically important trait. The most significant factor in how soon (or if) a cow rebreeds after giving birth is how fat she is on the day she calves. If a cow is fat when she calves, she has plenty of energy that she can immediately pull off her back to meet the demands of starting lactation, repairing her reproductive tract, and restarting her estrous cycle; the ability to convert surplus forage into immediately available stored energy in the form of body fat is the ruminants' secret weapon that allows them to meet the nutritional demands of their bodies even when forage is in short supply. To make use of this trait, two things must happen; the animal must have access to forage that is in considerable excess of her daily needs, and she must have the abilities required to harvest this excess and store it as body fat. If bred females with the proper genetic makeup have thirty days or more of unlimited access to high-quality green pasture before they give birth, they will breed back quickly, and the young will be born vigorous and with few complications. The excellent nutritional state of the dam is transferred over to the offspring and results in a strong calf with a healthy immune system that is able to overcome stresses and infections that would devastate a weaker calf born to a dam in nutritional stress.

This is the logical starting point for the prevention of the host of metabolic and infectious diseases that so often kill young calves. It doesn't make a lot of sense to gain fifty pounds of extra weight per live calf by spending the money, time, and suffering demanded by winter calving if you lose 10 percent of the calves born and have to spend time and money treating sick calves to keep mortality from going higher. Females that are thin when they give birth are often not able to take in enough energy, even on high-quality pasture or supplemental feed, to do all that needs to be done in a timely manner. It takes a lot of energy to repair the ravages of birth, properly feed the new young, and restart the estrous cycle quickly enough to stay on a twelve-month calving interval; energy stored as fat is immediately available for these needs. It is possible to maintain cows in the required nutritional state with heavy supplementation, but doing so increases both the cost of production and the amount of labor required. Fall or winter (dormant season) calving can make financial sense if ample, low-cost, cool-season forage is available, but it seldom makes sense if much supplementation is required.

If the breeding season is kept short, it is much easier and less costly to provide the proper nutrition for animals in the different stages of reproduction, because the whole herd goes from dry to wet in a short time. For most operations, the largest cash expense is winter feed, so it makes financial sense to keep the period when this is required to a minimum. If the herd contains animals in a wide range of reproductive states from dry to sucking big calves, with a single level of supplementation, some animals will always be either underfed or overfed. Feeding animals more than they need wastes money, and although keeping cows fat makes us happy, it doesn't help the health of the cows. Having all calves born in a short time period means that they can all be worked at the same time and weaned at the same time, saving stress and labor. Large groups of calves that are uniform in age will normally command a premium price. A short calving season also allows any cows that do not calve to be sorted off for sale while they are in good condition and without having to pregnancy test them.

Although it has already been discussed in the chapter on managing drought, the practice of overwintering weaned calves can play a large role in managing the financial aspect of later calving; the calves born on spring grass will be light in the fall, because they are young—but given the right genetics and some planning, this can be turned into an advantage. If two calves with similar genetics are weaned at six months and eight months, there will be little difference in the weights of the two calves coming off grass the following summer, and the younger, lighter calf will have cost less to winter.

When calves are weaned at seven to eight months of age, they have incurred something like 70 percent of the expense necessary to take them to heavy feeder weights. This is because they must pay not only the expense required to take them from birth to weaning but also the expense needed to maintain their dams for a year. The expense from weaning to feeder weight is only for what is required by the calf, so the cost per pound of this gain is much less. If the calves are weaned and given only the supplement needed to keep them healthy and growing normally, when spring flush comes, these cattle will be able to gain very rapidly and cheaply. This program allows a lot of flexibility in that the number of calves carried over can be varied to suit developments in weather and the market. It requires planning and some extra work, but some version of this program can usually be more profitable than selling at weaning. It works best if the cattle are moderately sized, moderate in milk production, early maturing, and able to store a lot of body fat—and if they are handled with good stockmanship and good grazing management but that, of course, is true of most programs.

There is another program that is simpler to operate and offers some real advantages in developing future breeding stock; if the cows are the right kind (easy fleshing, moderate sized, and moderate to low milk producers), and if pasture conditions are good, the calves can be left on the cows until they are ten to eleven months old. This is the system used by wild grazers such as bison and elk, and it provides the young animals with a great start in life. Many of the beef cows in use today give too much milk for too long and do not have enough ability to store energy as body fat to be able to nurse their calf this long without losing too much body condition. My ideal beef cow would hit the peak of her lactation about fifty days after birth with about one and a half gallons a day of high butterfat milk; this would drop off rapidly until she was giving less than a quart of milk at eight months, and she would gain weight on decent pasture for the last half of her lactation. Research has shown that after sixty days, calves from both high- and low-milk–producing cows gain at about the same rate; the calves from low milking cows develop into true ruminants earlier and make better gains on forage. Even a little milk is of real benefit as calves get older, because it supplies by-pass protein and other compounds needed by the growing calf. However, if the calf is fat, has a well-developed rumen, and forage is plentiful, milk is not needed for energy. Whether it is crop yields or weaning weights, many of our management decisions are influenced by the "bigger is better" philosophy; if you are hit with an overwhelming urge to go buy some big cows and feed them really well, take a deep breath, and look over the data on cow size and profitability in the chart in chapter 9.

CHAPTER 16

Adding Sheep or Goats?

There are real advantages to adding other species to a cattle-only operation if the available vegetation is suitable; many cattle operations could support a ewe or a nanny or both for every cow, with no competition for forage. The weeds eaten by sheep and the shrubs eaten by goats become assets rather than liabilities, and their consumption will increase both energy flow and the local nutrient cycle. The market for both sheep and goats is good and getting better; the U.S. imports both lamb and goat meat in large quantities. There is, however, a downside; sheep and goats are not "little cows," and most people who are new to sheep and goats will have a pretty steep learning curve to get up to speed. There may be capital costs in providing the fencing needed, and it is critical that whoever is caring for the animals understand their special needs.

Both sheep and goats originated in the drier regions of the world and to this day are best adapted to dry areas. To successfully manage sheep and goats in humid areas, it is necessary to understand how these animals react with their environment. Both sheep and goats are selective grazers rather than mass grazers like cattle or bison. They are physiologically adapted to consuming small amounts of high-quality feed rather than large amounts of low-quality feed. If possible, they will select only the young leaves and tender growing points of a wide variety of plants. Both species are equipped to metabolize plant toxins such as tannins to a much greater extent than are cattle and routinely select plants that mass grazers find repugnant. This adaptation is illustrated by the fact that goats have 30 percent more liver, measured as percent of body weight, than cattle have. Sheep and goats are adapted to utilizing plants different from those used by cattle and this trait gives them great value in grazing management. It must be remembered, however, that they, like all grazing animals, require forage that is digestible and that contains a proper blend of protein, energy, and minerals. Far from the popular belief that "goats can live on anything" is the fact that goats are the most nutritionally demanding of all domestic animals. Goats and hair sheep can utilize and need large amounts of browse (the leaves, buds, and tender growing points of woody plants), but this does not mean that they can thrive or even exist on a diet made up solely of mature oak leaves. Forage quality is primarily a function of age, and goats can no more utilize senescent tree leaves than cattle can thrive on wheat straw. Attempting to hold goats on brush long enough to kill the brush is a guaranteed recipe for disaster and has been proven so by several ill-considered university-sponsored research projects. It is quite possible to kill the species of quality browse plants by grazing the plants on a rotational basis using too short a recovery period; a more logical approach would be to consider the browse a resource and graze it correctly so that the desired balance of vegetation is achieved.

Pasture composed mainly of grass that would be very high quality for cattle is less than adequate for sheep and totally unsuitable for goats. These species need the additional minerals and unique plant compounds found in the tap-rooted forbs and browse that is their preferred feed. Sheep and goats may exist on a diet of Bermuda or fescue grass, but they will be nutritionally deprived and under constant stress from parasites and disease. They will also be very hard to keep home, because they are smart enough to know that they are not receiving what they need and will make every attempt to go somewhere to find the needed nutrition.

Stocking the land with a mixture of species in proportion with the vegetation available for each species is a proven technique to improve the productivity and stability of the land and the profitability of the operation. When all portions of the available vegetation are being utilized correctly, the amounts of various nutrients being cycled is greater, the rate of cycling is faster, and much more energy flows through the entire system. On an area of mixed grass, forbs, and browse, it is possible to double the amount of protein produced per acre by stocking a mix of cattle, sheep, and goats as opposed to stocking only one species.

Vegetation is sparse in dry regions, so animals dependent on the vegetation must range over larger areas than would be required in a more lush environment. Sheep and goats evolved in situations where internal parasites were at a disadvantage (dry conditions and low animal concentration); without the need, neither sheep nor goats developed a high degree of genetic resistance to internal parasites. At the same time, many of the internal parasites that infect sheep and goats developed traits that allowed them to survive with these disadvantages. Most of these parasites produce very large numbers of offspring, so even if only a small percentage survives the harsh conditions, the species will survive. The parasites also developed the ability to enter a state of reduced metabolic activity to survive longer outside the body of their host animal, thus having a better chance of obtaining a host. When sheep and goats are moved into a humid area, the abundant forage allows for a greatly increased concentration of animals and of their parasites. This concentration, along with the more favorable conditions for parasite survival, brings about tremendous increases in parasite infestation. Internal parasites and the resulting loss of vigor are the main problem of sheep and goats in humid regions. Heavily parasitized animals are much more susceptible to predators and to disease, which rank as the second- and third-largest problems in sheep and goat husbandry.

It requires effort on several fronts to maintain parasite loads at acceptable levels anywhere, but a multifaceted approach is an absolute necessity in humid regions. The conventional method of dealing with internal parasites has been, for many years, to dose the animals with some sort of toxin to kill the parasites after the parasites have become established. Many different compounds have been used over the years, from a copper sulfate and nicotine sulfate mixture to phenothiazine, to the made for the purpose chemicals created and in common use today. The effectiveness of these materials varies widely, as does their toxicity to man, beast, and the environment. Such things as tetrachlorethylene, carbon tetrachloride, phenothiazine, and nicotine sulfate are no longer used because of their toxicity. The made-for-the-purpose vermifuges are generally less overtly toxic and are effective in killing parasites at least until they have been in use long enough for parasites to develop genetic resistance to them. In some parts of the world, the economically important internal parasites are resistant to all commercially available antihelmintics, and there are no effective materials available to kill parasites in sheep or goats. Even when effective low-toxicity compounds are available, it is not possible to control internal parasites solely through antihelmintics in any but the most arid areas. When an effective de-worming agent kills all, or nearly all, of the parasites in an animal, antibody (material produced by the host animal to combat parasites) production stops or greatly decreases. When the level of protective antibodies in the animal drops and the animal is again exposed to infective parasite larvae, the reduced antibody level allows most of the larvae to become established, and the animal suffers a severe parasite infection. Another downside of most chemical wormers is their deleterious effect on dung beetles, earthworms, and other beneficial organisms that play vital roles in nutrient cycling and in the reduction of parasite levels. Control of internal parasites can be achieved without toxins by adopting a management regimen that attacks the problem on a number of different fronts. An effective program for dealing with internal parasites consists of three main parts; promotion of animal health (and thus healthy immune systems), selection for genetic resistance, and reduction of parasite levels in the environment. Intervention with antihelmintics should be practiced only in crisis situations and not as a routine part of animal husbandry.

The key to producing healthy animals with vigorous immune systems is to make certain that all of their needs (physical, nutritional, and psychological) are met. The vast majority of the disease and parasite problems that afflict our animals are caused by mismanagement on our part. It is not possible to maintain livestock in an environment that is totally free of pest organisms; infective agents are always present, but healthy animals are able to resist most infection.

Stress in any form weakens an animal's ability to combat the organisms that cause disease and parasitism. The source of stress can be overt, as with being forced to live in their own filth, rough handling, severe climatic conditions, lack of feed, poor-quality water, or dusty conditions, but it can also be caused by more subtle factors, such as nutritional deficiencies due to the wrong type of feed, disruption of the social order, or by the requirement of behavior foreign to the animals' nature. Goats and sheep are able to survive under poor feed conditions, because they have the ability to selectively gather the very high-quality small leaves and growing points that are not available in sufficient quantity to mass grazers such as cattle. This ability is the source of their reputation as destroyers of rangeland, and they are very destructive when kept in one area for too long. The main cause of their damage to plants is the ability of sheep and goats to selectively graze growing points before they have the opportunity to become functional and so prevent the plants from renewing their foliage. When the high quality forage is gone, the animals are in nutritional stress and subject to diseases and parasites long before serious damage is done to the vegetation. Both sheep and goats are selective grazers that take in relatively small amounts of feed and so require a diet that is higher in quality than that required by cattle. In the proper conditions, both species satisfy this need by selectively grazing the growing points of mainly forbs and shrubs with some use of immature grasses and seedheads. If they are forced to subsist on grass, with its lower nutrient content, or on mature shrub leaves, they soon become nutritionally deprived and lose vigor. In an effort to gain the needed nutrition, they will graze closer to the ground and thus pick up heavier loads of parasites. To thrive, both sheep and goats require access to a diverse and healthy mixture of plants, forbs, shrubs, and grasses from which to select their preferred diet. Supplementation with proper amounts of minerals, protein, and energy can help, but there is no substitute for quality pasturage. Recent research has shown that adding kelp to the mineral mixture has real benefits in stimulating immune system health, and some producers feel that adding garlic powder, unprocessed diatomaceous earth, and dried apple-cider vinegar serves the same purpose.

Stress from psychological factors can be just as damaging as stress from physical sources. Both sheep and goats have highly developed herding instincts and become stressed when separated from the flock. This trait is so strong that mothers will often leave their newborn offspring rather than be left behind when the flock moves. No herd animal should ever be isolated from other animals, and managing these animals so that the herd remains the social unit provides real health benefits. This is particularly important at lambing or kidding time, and the grazing should be planned so that the animals have access to the quantity and quality of feed required without the stress of frequent moves or supplemental feeding. If animals can be kept content and stress free, many of the maladies that afflict them will disappear. An excellent practice is to set aside an area with enough quality forage to maintain the females for at least the major portion of the lambing/kidding season and to not graze this area with sheep or goats except this one time a year; it can and should be included in the grazing rotation for other species such as cattle, horses, or bison during the rest of the year.

Selection pressure is one of the most powerful tools available to the livestock breeder. It is fairly easy to breed animals with higher weaning weights, more or less wool, smaller mature size, or one of many other traits simply by mating "the best to the best." To do this, it is necessary to determine to what extent various animals demonstrate the desired trait. If calves are creep fed from the time they are old enough to eat, it will be difficult to determine which cows are deficient in milk production. The same is true of parasite resistance in sheep or goats that are managed under a routine worming program. If routine worming is stopped, it will be apparent which animals have genetic resistance to parasites and which do not. If those animals showing signs of parasitism are culled and breeding stock is selected from those not showing damage, in a few generations, it is possible to greatly increase the genetic resistance in a flock. The same technique works on other traits, such as foot rot, balloon teat, and long toes. Even traits that show low estimates of hereditability improve dramatically in a population that is subjected to intense selection pressure for or against these traits. The advent of antibiotics and chemical antihelmintics has made it possible to save and even breed animals that in earlier times would not have survived due to genetic weakness. Good husbandry dictates that we use the materials necessary to prevent death in these animals but that we not perpetuate the poor genetics by allowing these animals to reproduce.

If we wish to manage to reduce the number of parasites present in an environment, we must understand the life cycle of the pest and attack it at its weak points. The majority of parasites are most vulnerable during that period when they are outside the body of the host animal. Most parasite eggs are shed in the spring when temperatures are warm and moisture is most plentiful, so during this period, management can be used to reduce parasite loads. The organisms are limited in the time that they can survive outside a host by the limits of their biological clocks, their energy reserves, predation by other organisms, and climatic conditions. The number of viable parasites on an area decreases rapidly with every day beyond what is required for the parasite to hatch and grow into the infective stage. One of the most effective means of reducing parasite numbers is to remove all susceptible animals from the area until most parasites have expired. As most parasites are host specific, this method is even more effective if the area can be grazed with nonsusceptible animals while it is empty of susceptible animals. If an area is empty of sheep and goats for 60–90 days during warm weather and grazed with cattle during that time, it will have very few sheep and goat parasites. The reverse is also true: sheep and goats help to reduce the number of cattle parasites.

Tillage also greatly reduces the number of viable parasites on an area and can be used to prepare a clean area with high-quality forage on which to wean lambs or lamb out a set of ewes. The ability to control which areas are grazed and at what periods is a necessity in all livestock management, but it is especially critical with sheep and goats. Ten to twelve paddocks through which sheep or goats can be rotated is a minimum number if parasite control is to be achieved primarily through good hygiene. Special care should be taken to reduce common areas such as water lots, lanes, or any other areas that are used continuously or frequently by the animals, as these serve as reservoirs for disease and parasite organisms. With the ability to practice time-controlled grazing, the manager can control the height of the forage sward so that animals are not allowed to graze forage shorter than 3–4 inches before being moved to fresh pasture. This maximizes nutritional quality for the animals while promoting vigor in the forage and also reduces parasite infection, as most worm larvae do not crawl up higher than two inches on forage. It is critical for this management that sufficient quality growing time be provided so that the forages being grazed make complete recovery before being grazed again. A secondary benefit to good, time-controlled grazing is an increase in soil life, which is the basis of soil productivity and also dramatically reduces the population of pest organisms that spend a part of their life cycle in or on the soil. In humid areas, it will be extremely hard to control internal parasites in sheep and goats without being able to plan the grazing sequence and timing through multiple paddocks and without the ability to practice multiple species grazing.

The grazing techniques, which reduce parasite levels, also help to eliminate disease problems such as foot rot, overeating disease, lamb scours, and any other malady that can be prevented by good nutrition, healthy immune systems, and clean pasture. The same techniques also reduce the incidence of metabolic upsets such as bloat, milk fever, and pregnancy toxemia by supplying forage of the proper kind and age. Poisonous plants also become much less of a problem, as the animals' needs are being met and they are much less likely to eat highly toxic plants. Hygiene is always important but is doubly so in humid areas where disease and parasite organisms can survive for long periods of time. It is critical to provide clean water that cannot be contaminated by manure or urine; in addition, barnyards, lanes, and shelter sheds should be fenced out except when actually in use. Changing the working area where lambs and kids are marked by using portable pens will avoid the buildup of pathogens in the soil of the working pens and help prevent diseases such as tetanus, blackleg, polyarthritis, and CLA.

Setting the lambing/kidding season to coincide with good pasturage and good weather goes a long way in preventing expense and loss of animals. Females that receive green pasture prior to parturition will produce and save more and healthier offspring; there is no feed of any kind that can provide the nutritional quality and balance of green pasture.

Ewes and nannies are most susceptible to internal parasite infection when their hormonal levels change just before and after lambing. If ewes can be moved to parasite-free pasture the week before lambing starts, infection levels can be reduced. The infection rate can be further reduced if the paddocks being used contain forage with high tannin content. Tannins prevent internal parasites from implanting in the animals gut wall; this also works for cattle. Some common forage plants that contain tannin are sericea lespedeza, common lespedeza, arrowleaf clover, Berseem clover, birdsfoot trefoil, crown vetch, and sainfoin.

Setting the breeding season so that young are born in the winter and thus big enough to utilize spring growth and wean off at heavier weights is seldom a good strategy. The females will require supplementation of some sort, death loss and sickness of both dams and young will be high, and predation is apt to be high, as food for predators is scarce. In west Texas, loss to eagles can be severe during winter lambing, but it drops to nothing during spring lambing. In addition, one ice storm in the middle of lambing season can reverse any possible economic advantage for winter lambing for many years. If spring-born lambs are too small to meet the market requirements, providing supplemental feed for them usually makes more economic sense than winter lambing. One of the real advantages of humid area production is that the grazing can often be planned so that lambs can be weaned onto high-quality pasture and finished for slaughter on pasture alone. The secret to success is planning; humid areas offer advantages that can be exploited and disadvantages that can be overcome by a well-designed production program.

Sheep and Goat Predator Control

Shepherds have been trying, with varying degrees of success, to protect their flocks from predators for as long as there have been domesticated animals. For the most part, these efforts have employed two main techniques: (1) guard the livestock, and (2) kill the predators. In the sheep and goat country of the western United States, an intense program of predator eradication was undertaken beginning in the late 1940s. Trapping and poison had been used for many years in this area, but nothing like the widespread use of long-lasting systemic poisons that virtually eliminated the larger predators such as coyotes, wolves, bears, cougars, bobcats, eagles, and foxes over a large area. The program also greatly reduced the numbers of the smaller predators, such as skunks, raccoons, hawks, and owls as well as those of carrion eaters such as opossums, vultures, and crows. The program made the sheep and goats in the area safe from predators but at a tremendous cost; the removal of predators allowed the populations of prey animals to explode and consume the vegetation needed by the livestock. In the early 1950s, there were areas in west Texas where jackrabbits consumed all of the grass and forbs as fast as it grew, leaving nothing for livestock. In these same areas, packrats were so numerous that they stripped the bark off of and killed the valuable browse plant species relied upon by goats and deer. Many in the livestock industry did not wish to recognize it, but the value of predators in nature's scheme had been forcefully demonstrated. Although somewhat harder to recognize, also demonstrated was the danger we incur when we manage against what we don't want instead of for what we do what. If our management focus is on producing healthy soil, healthy plants, and healthy animals, many of our problems with disease, parasites, predators, and weeds will automatically be greatly reduced.

Predators in sheep country are like cold viruses in a schoolyard; they are always around and always ready for a victim. Holding predator losses to a reasonable level requires a multifaceted approach that addresses a number of factors, including the type and concentration of predators, the experience of predators as stock killers, the availability of prey species other than sheep and goats, and the attractiveness of the livestock to the predators. If there are experienced stock killers present that have been accustomed to making their living by killing sheep, very little can be done except to remove the killers—be they coyotes, bobcats, or whatever. By the same token, if predators are present and are not killing stock, these animals, because they are highly territorial, become a powerful deterrent to keep other predators that might be killers from coming into the area. As strange as it seems, a coyote that is not killing stock is the sheepman's best friend; not only does he control rabbits and rats that compete with the sheep for feed, but he acts as a guard to keep other coyotes out.

Normally, predators take sheep and goats as prey because the livestock is more readily available than their normal prey or because the livestock is an easier kill. If lambs or kids are being born in the dead of winter when there are no baby rats and rabbits around, the new lambs and kids become a powerful temptation to every predator in the area. The sloppy stockman who allows young to be born at all seasons and thus always has young animals left behind the flock will soon discover that he has created stock killers. The same is true of stockmen who allow their animals to become scattered at low stock density over a large area due to sparse graze or other reasons. An animal away from the herd is much more likely to be taken by a predator than the animal that stays with the herd.

Penning at night can be an effective predator deterrent, but almost as good results can be achieved by training the livestock to come together at night on a bed-ground in a compact mass. These bed-grounds can be changed regularly to prevent parasite and disease buildup and to prevent land degradation.

Sheep and goats die and allowing the carcasses to be consumed by predators or even by carrion eaters is to invite these animals to become stock killers. Animals such as black vultures that are not normally killers can be corrupted by the sequence of eating dead lambs to eating afterbirth to attacking the female and baby during the birth process.

Sloppy trapping and poor shooting have created many stock killers, but even good predator control programs are valuable only when they are targeted at stock killers. A crippled predator that lacks mobility is much more apt to become a stock killer, because he always knows where the sheep are and doesn't have to travel to find prey.

Predator density is controlled largely by the size of the food supply. When food is scarce, litter sizes decrease, and fewer young survive; when food is plentiful, more young are born, and a larger percentage of them reach maturity. When predator numbers are reduced, the prey animal populations explode, and more predators will come in to exploit the increased food supply.

Using guard dogs can be one of the best ways to protect sheep and goats; a secondary reason guard dogs are effective is that they reduce the prey animal populations by hunting and thus make the area less desirable to predators. Guard dogs best protect livestock by controlling an area centered on the location of their charges. If the dogs are fenced in with the livestock, they are at a disadvantage, because they cannot patrol the surrounding area and intercept predators before they reach the livestock. There have been instances in which coyotes have been seen to dive between the wires of an electric fence, grab a lamb, and dive back through the fence before dogs with the sheep could react. The predators soon learn when dogs are fenced in and unable to give chase, and they make use of this information. Dogs are the most widely used guard animals, but llamas, donkeys, and horses have been used successfully as well. To be effective, these animals must be bonded with the sheep or goats and, unlike dogs, work best when only a single animal is used per flock. They are most useful when paddock sizes are small and fairly open. Some people have had success bonding cattle with sheep or goats and running the two together, but this practice is hard to manage when grazing rotationally or with large numbers of animals. Although possible, it requires elaborate fencing and constant vigilance to fence predators away from livestock, and it is seldom practical to do this on large areas.

Conclusions

Ranching both as an industry and as a lifestyle changed very little for many years; in the last fifty years, changes have been rapid and dramatic. Some of these changes have been positive and some have only appeared to be positive. The following describes what I consider to be some of the major problems facing ranching and some suggestions on how they be addressed.

Problem: Poorly adapted animals

The rapid expansion of feedlots and acceptance by consumers of corn fed beef as the standard changed the dynamics of the beef cattle industry. From playing a relatively small role, by the 1960s, the feedlot industry became a dominant factor in determining the type of cattle that would be produced and how they would be produced. Government subsidies created surpluses of cheap grain and the cattle feeders wanted big cattle that could turn a lot of corn into beef without getting fat. The packers also liked big cattle because a lot of their costs occurred on a per head basis and big carcasses were cheaper (per pound) to process. Cattle producers responded to the market demands and cattle got bigger, slower maturing, and much leaner. With larger cattle and a larger percentage of cattle being fed grain, the amount of meat produced per head slaughtered rose dramatically; the industry now needed fewer mother cows to produce the same or greater amount of meat. There were fewer of these new cows but they changed completely the way that most ranchers operated. By selecting for bigger carcasses with less backfat ranchers gave up much of the ability to produce cattle solely on grass. The new style animals required much more nutritional supplementation and were reproductively less efficient. By producing what "the trade demands", the cow calf producer gave up a lot of their chief economic benefit, converting cheap forage into beef. The need to provide the greater nutritional needs of the big cattle led to a mechanization of ranching and a large increase in the cost of production. Where in the past most ranchers spent money primarily for grass and animals, machinery, fuel, and labor was required to provide for the needs of the high maintenance cattle. Ranching became a capital intensive endeavor with many ranchers putting more money at risk (purchasing inputs) than was justified by the potential for profit. Ranching went from being a wealth producing endeavor of the first order to being a subsistence life style for many ranchers.

Remedy

Ranchers have turned into farmers trying to provide for the needs of cattle that are basically unsuited to survive on grass. Most of today's beef cows are too big, too slow maturing, give too much milk, have too little backfat, and have too little rumen capacity to be successful ruminants.

The cattle needed for success in a grass based program will be: moderate sized with thick and wide bodies, early maturing, moderate to low milking, easy fleshing, docile, fertile, genetically tender, and have the large rumens and wide mouths needed to gather and process large amounts of forage. These animals will be consistently more profitable than the over sized gut less wonders beloved by the feed yards. By adopting preventive veterinary medicine practices such as vaccines, routine worming, and antibiotics in feed, we have allowed animals that are poorly suited to our local environment to reproduce. Good husbandry demands that we treat sick animals but it is poor practice to allow them to reproduce. Genetic selection can be used to reduce or eliminate maladies from bad feet and balloon teat to horn fly and internal parasite infestation. Culling and good hygiene and nutrition through grazing management can eliminate much of the need to spend money on veterinary products and treatments.

Problem: Over reliance on technology

Agricultural technology exploded after World War Two bringing greatly increased production but with mixed results as to the financial well being and quality of life of farmers and ranchers. I was an early and complete advocate of the adoption and use of agricultural technology; if Texas A&M or OSU recommended a practice, I did it in spades. As early as 1960, I was using management similar to the *best practices management* that is in vogue today. It would take more than ten years for me to realize that this type management has basic flaws that make it extremely dangerous. Many of the practices were especially dangerous because their deleterious effects were not immediately apparent. By pursuing high production (which I got) above all else through technology, I created a situation that was unhealthy on all fronts. Our soil lost ability to take in and hold water as we burned the organic matter content out of it with tillage and nitrogen fertilizer; when the organic matter was gone, soil life (the force that drives soil health and productivity) decreased dramatically. It took increasing amounts of fertilizers to maintain production and signs of trace mineral deficiencies begin to appear in plants and in the animals feeding on the plants. These problems did not appear overnight and I was slow to recognize their cause when they did appear. I did not (for a long time) connect things like increased weed pressure, reduced animal fertility, and increased animal health problems to my management practices. This failure points to a major fault in the manner that technology is used and evaluated; the focus is too often on a very narrow set of readily measurable factors. We look at pounds of forage grown per acre and the out of pocket cost of that forage but not at whether or not the increased amount of forage is the best use of the money spent to achieve it or even if more forage is a net positive for the whole operation. We are also frequently rather shallow in our analyses of the situations we encounter; when a problem is recognized, our first reaction tends to be how to kill the weeds or treat the disease rather than to determine why the weeds or disease became a problem. We tend to use technology as a crutch to allow us to limp around a malady rather than expend the thought and energy necessary to understand and cure the root problem of which the malady is a symptom.

Remedy

In the 1960s it was common to hear, "The solution to problems of technology is more technology." From the standpoint of agriculture, this is definitely not true; the solution to problems of agricultural technology will be found in better understanding of all of the natural sciences: plant and animal physiology, soil science, ecology, and how the relationships between these sciences impact the natural world. We do not need stronger pesticides; we need to understand why we have pests. Agriculture is a biological process with very few problems that are curable by quick and simple solutions. Many of the "cures" touted in the trade magazines and even recommended by agricultural experts not only do not solve the perceived problem but actually intensify the ill while reducing financial health. Some of these remedies change the physiology of the soil-plant-animal complex so severely that the system becomes dependant on continued use of the product, often at ever increasing rates, just to maintain current production. Insecticides that kill beneficial insects as well as pests and soluble nitrogen fertilizers that destroy soil life are examples. Nature uses biodiversity (complexity of life forms) to increase efficiency of resource usage and also to prevent organisms from building population numbers so high that they become pests to the other members of the soil – plant – animal complex. By managing for what we want (healthy soils, plants, and animals) rather than to kill what we don't want (coyotes, brush, horn flies), we build biodiversity and pest organisms of all kinds become less important; improving biodiversity also increases the amount of solar energy converted to biological energy and enhances both production and stability of the local soil – plant – animal complex.

Problem: Poor understanding of relationships between forage and grazing animals

A very real problem and one that we have not handled well is the steady decrease in health of our natural grasslands; several years ago Texas A&M published data showing that the carrying capacity of Texas ranges had decreased an average of five percent per year since 1900. If it achieves nothing else, I hope that this book convinces some people that the degradation of grasslands can be halted and reversed and that it does not take heroic measures and loads of money to do so. Most "range improvement" practices are addressed at treating symptoms of poor range health (brush encroachment, species loss, poor water infiltration, etc.) rather than the causes of poor range health. Many of these practices (besides financial cost) make the situation worse in the long term by reducing biodiversity.

Remedy

I am convinced that most of the damage done to native grasslands comes from two main causes: continuous grazing (continuous exposure to defoliation) and holding animals (by substitution feeding) on land that can no longer feed and water them. Both of these practices cause a reduction in the health of the ecological processes (water cycle, mineral cycle, energy flow, and biological succession) and the damage from both can be corrected by planned time controlled grazing; people all over the world have and are using good grazing management to reverse range degradation. In the process they are finding better health for their range, their animals, their finances, and themselves. An excellent definition of good grazing management would be management that promotes the health of all parts of the whole.

Problem: Poorly defined goals

For a number of years I worked hard to increase the weaning and yearling weights of our cattle; I tagged each calf at birth keying it to its dam and weighed each calf individually at weaning. I used this information to select the heaviest heifers to go back into the cowherd. I started an A I program using some of the fastest growing bulls in the Angus breed. When I started, our cows weighted 1000 to 1100 pounds and our calves 210 day weaning weights were around 450 pounds. After ten years or so of this selection plus some other changes like moving to winter calving, our calves weaned at over 100 pounds more and our cows now weighed 1300 to 1400 pounds. I set out to raise the weaning and yearling weights of our calves and I made this happen. I raised the weight of our cattle on the theory that more pounds of calf at weaning would mean more pounds to sell and more profit; this did not happen. Larger cows meant that I could not run as many cows and the larger cows were not as fertile as the smaller cattle so that I wound up selling fewer pounds of calves. In addition, the bigger calves brought less per pound while their dams cost more to maintain. I achieved my goal but it was the wrong goal. This was not a good period in my career; I was doing things right according to conventional wisdom but I was getting bad results on all fronts. The ranch was losing money and the harder I worked, the worse it got; this is probably the period when my kids decided that they had no interest in becoming ranchers. In 1974 a bad thing happened that turned out to be a good thing for Davis Ranch; the cattle market broke and we were left with the choice of reducing expense or losing the ranch. We started cutting costs and rearranging our priorities from production to profit and good things began to happen. We saved the ranch and eventually developed a management program that was more profitable, more stable and a whole lot more enjoyable.

Remedy

It is very easy to get caught up in the practices and the processes of running an operation of any kind and lose sight of the purpose of the operation. Ranching is a business that needs to be run like a business but it is also a life style. First decide what is important to you and yours and work from that point to plan the management that will give you what is important. I found that the closer I could model my management to the way the natural world works, the better I liked it. It is amazing how little labor and how little money it takes to run a grazing operation (once the infrastructure is in place) when animals are adapted to the management being used and production practices are based on natural cycles. I found that I really didn't miss making hay all summer under a blazing sun or feeding it all winter in a freezing wind. That, however, is just me; you will have to find what pleases you.

Ranching is a business but don't forget the purpose of having a business

---

Made in the USA
Lexington, KY
23 December 2015